SCIENTIFIC KNOWLEDGE

Scientific Knowledge

A Sociological Analysis

Barry Barnes, David Bloor & John Henry

The University of Chicago Press

The University of Chicago Press, Chicago 60637

Athlone Press, London, NW11 7SG

© Barry Barnes, David Bloor, John Henry 1996

All rights reserved. Published 1996

Printed in Great Britain

05 04 03 02 01 00 99 98 97 96   1 2 3 4 5 6

ISBN: 0-226-03730-4 (cloth)

ISBN: 0-226-03731-2 (paperback)

Library of Congress Cataloging-in-Publication Data

Barnes, Barry.
  Scientific knowledge : a sociological analysis / Barry
Barnes, David Bloor, and John Henry.
      p.   cm.
  Includes bibliographical references and index.
  1. Science--Philosophy.   2. Science--Psychology.
I. Bloor, David.   II. Henry, John.   III. Title.
Q175.B2184   1996
306.4'5--dc20   95-45231
                                                  CIP

This book is printed on acid-free paper.

We dedicate this book to

David Brock
Philippe Chavot
Eric Francoeur
Rhodri Hayward
Rebecca Hodgson
Perry O'Donovan
Seamus Prior
Bart Simon

Contents

Introduction

Our objective in this book is to state precisely and clearly where and why sociological analysis is necessary in the understanding of scientific knowledge. Our main method is to present historical case studies. We then show how sociological analysis applies in these cases, and how it is an essential complement to even the most insightful interpretations derived from other perspectives. In establishing this last point we introduce important work on science drawn from psychology and philosophy. We try to present this non-sociological material in the most favourable light so that, rather than overestimating the power of the sociological approach, readers will appreciate that sociology makes a necessary contribution to a larger project of obtaining an understanding of science.

We see the sociology of scientific knowledge as part of the project of science itself, an attempt to understand science in the idiom of science. Other sociologists have attempted to develop perspectives on science using approaches which are uncharacteristic of science, and which do not accept or rely upon its methodological conventions or its accepted cosmology. We ourselves honour science by imitation: in our study of science we try to emulate its own matter-of-fact, non-evaluative approach. Ironically, some scientists and philosophers have assumed that since we neither praise nor defend science our objective must be to subvert it. They have failed to understand that for a social scientist to seek to justify science would be to deviate from its own non-evaluative precepts. One cannot simultaneously adopt a scientific approach and celebrate it.

We begin our study with an examination of the reliance placed by science on observation and experience. A number of psychologists and philosophers have argued that observation is theory dependent. They say our minds actively create part of what we perceive, and do so in a manner that expresses the theoretical presuppositions of the perceiver. This view would seem to have clear advantages for a sociological approach to the acquisition of knowledge. Drawing upon recent empirical and theoretical work, however, we argue in Chapter 1 that there is no *need* for sociologists of knowledge to take this view. It is perfectly possible and empirically plausible to accept the claim that perception is to a large extent 'modular' – that is, isolated from other components of our cognition, and only influenced by them in a limited way. This in no way rules out a sociological analysis.

The reason is that the raw material of our experiences must be transformed into observation reports before it can begin to enter into scientific knowledge. In the second chapter, therefore, we look at ways in which a consensus is reached as to the correct interpretation of these experiences. We proceed by giving a sociological analysis of R.A. Millikan's experimental attempts to measure the charge on the electron. Basing our account on Gerald Holton's detailed examination of Millikan's experiments and the records of them left in laboratory notebooks, we argue that Millikan's interpretation of his results depended upon an available tradition in his subculture. This has profound implications for our understanding of the interaction between nature and culture.

In Chapter 3 we look at the character of classification and try to formulate some general conclusions about how words relate to the world. We begin by considering how classification would proceed if it were based wholly on assertions of resemblance and identity between ostensively given examples of a class of thing. This is a simplification, and later discussion introduces the relevant complexities, but it serves to introduce the view of meaning called 'sociological finitism'. The essential point is that our classifications are always underdetermined by the promptings of experience or by previous acts of classification. Each new application of a term is sociologically problematic. A finitist account thus emphasizes the conventional character and sociological interest of classificatory activity, such as the fact that every act of classification has the form of a judgement, every act changes the basis for the next act, every act is defeasible and revisable, and every act involves reference, not just to the 'meaning' of the term applied, but also to the 'meaning' of all the other terms currently accepted for use in the context.

Scientific classification does not, however, treat things as conventional clusters of diverse particulars: it treats things as falling into *natural kinds* whose members can be thought of as essentially identical to each other, or as identical in certain important respects, for example as electrons, as a particular chemical element, or as members of a biological species.

Objects which are members of natural kinds are things about which true or false statements can be made, and to which fully general scientific laws can be applied. Chapter 4 deals with the insistently realistic idiom in which scientists treat the objects referred to in their laws and theories.

Some sociologists see realism as a style or assumption that is to be opposed. Our view is that it can be understood and illuminated by the finitist account already developed. First, scientists' talk of essential identity is clearly not a simple response to empirical prompting. It must be thought of as a preferred strategy of a group of scientists. Second, the 'theoretical discourse' of scientists cannot be seen as deriving from a fixed set of theoretical statements with determinate meanings from which significance for particular cases emerges. It is all too easy to talk of Dalton's theory, or Newton's theory or Mendel's theory, and conceive of them as sets of statements with fixed meanings. Once these theories are addressed as historical phenomena this approach falls apart. As would be predicted by finitism, it is extremely difficult to specify what a theory is as an historically situated entity, and it is quite impossible to identify it as a set of statements. It is better to think of a scientific theory as an evolving institution. Illustration is provided by appeal to historical studies of 'Mendel's theory'.

The first four chapters therefore show the role of tradition, convention, consensus, and the social processes whereby these are upheld or dismantled in establishing and sustaining knowledge. Chapter 5 shifts the focus and seeks to show in a more synoptic way what sociological study can add to our understanding of scientific knowledge and scientific practice. One concern here is to show that scientific research is conducted with reference to goals and interests, and that both research and the evaluation of research must be understood as intrinsically goal-oriented activities. Two major examples are discussed: the role of chemical arguments in the legal disputes concerning a patent protecting the manufacture of red dyes derived from aniline, and the interpretation of experiments in which fermentation was first held to take place outside a living cell.

Chapter 6 is concerned with how scientists defend the intellectual territory over which they wish to exert control, with how they demarcate science from non-science in order to present a credible image of specialist expertise and intellectual authority. A consideration of the origins of the 'experimental method' shows that our current image of science has been carefully forged by natural philosophers and scientists for specific practical purposes, and that further development and elaboration of the image built up over the course of two centuries or more continues at the present time.

Our final chapter looks at mathematics and mathematical reasoning. It is particularly important that we include this topic in our remit since mathematics has an important role in the natural sciences, and it is often claimed that mathematical knowledge defies analysis in sociological terms. We show that, on the contrary, sociological considerations have an essential role in understanding the status of the simple but nonetheless fundamental proposition that $2 + 2 = 4$. We go on to provide a general discussion of the character of proofs in mathematics, and to indicate what we take to be the proper sociological response to such achievements.

The text is the responsibility of the authors collectively, and the outcome of their working together over a considerable period. But we do not believe in writing by committee, and have drafted the various chapters individually, taking account of our different interests and competences. If there is some diversity of style from chapter to chapter this will no doubt be its source. If there is less than complete consistency of argument this could well have occurred for the same reason, but our intuition is that any such inconsistencies are unlikely to be major ones.

The book's particular form derives from its origin as an introductory course for graduate students. We hope that it will be of some value to others, from their final undergraduate year onwards, seeking a text in the sociology of scientific knowledge. Since there is no barrier in terms of vocabulary or presumed background knowledge of sociology, and since the text concentrates on basic issues, anyone with an interest in history or philosophy of science, or cognate fields, might also find the book of value as a way of relating sociological approaches to their own. Conversely, the case study materials do not discuss scientific knowledge in a way which demands any special training, so the book should be accessible to those with an interest in the sociology of knowledge and culture as traditionally studied. It should be remembered, however, that few of the historical case studies we cite have been written in order to advance the claims of sociologists of scientific knowledge. A number of the historians whose work we have borrowed may well hold a more realist or a more positivist view of scientific knowledge than that which is presented here, but they have produced exemplary empirical studies, and these are always grist to the mill of theoretical perspectives like that of the sociology of knowledge.

Although this is a demanding introduction to our subject, it is nonetheless an introduction. It concentrates on the basic foundations from which an understanding of more elaborate ideas may grow. Hence it gives little prominence to currently much debated issues like relativism, reflexivity and self-reference, the role of new literary forms and post-modernist 'deconstruction' – issues which are best approached after some extended consideration of simpler matters. Nor is there any attempt to survey the whole range of views and perspectives currently to be found in our field. When the sociology of scientific knowledge first got under way over twenty years ago, there was general agreement about its basic approach. The concern at that time was mainly to oppose the arguments of rationalist philosophers who wished to treat science as a unique form of human activity, one which required no empirical understanding other than that implied by describing it as rational. Now, however, differences of view in sociology of science are every bit as extreme and far reaching as those in philosophy of science. It is no longer possible either to give an account of 'the approach' established in the field, or a properly detailed account of the many and varied approaches actually to be found. It must be remembered therefore that the 'sociological analysis' of the book's title is our own. If it is occasionally referred to as 'the' sociological approach, the reader is asked to indulge us, or to look to our notes and references for ways into alternative perspectives.

Acknowledgements

We should like to express our gratitude to our students at Edinburgh University, particularly those who took the MSc course in the Sociology of Scientific Knowledge upon which this book is based. We are also extremely indebted to the various sociologists, philosophers and historians of science whose work we have drawn upon in the writing of this book. Two (anonymous) publishers' readers made a number of very valuable and helpful observations, and we have done our best to meet their points.

Finally we wish to thank Carole Tansley, for typing and retyping numerous drafts not only without complaint but also with great efficiency and good humour, and Brian Southam of The Athlone Press for his forbearance when we missed several deadlines.

Edinburgh 1995
Barry Barnes
David Bloor
John Henry

CHAPTER 1

Observation and Experience

There are a number of reasons for beginning an account of knowledge with the topic of observation. Our aim is to build up a description of knowledge as a natural phenomenon. No such account could fail to attend to the fact that the inputs from our sense organs act as an important stimulus to changes in our beliefs, and the formation of new beliefs. That doesn't necessarily mean that 'observation' is correctly identified with the deliverances of our sense organs – that may be only part of the story – but it is a good reason for attending to the role it plays in knowledge. Observation is rightly thought of as a channel or inlet through which the material world around us makes its presence felt. To understand observation is therefore to understand the role played by the reality of our environment in the formation of beliefs about it. All too often the charge is made that sociologists of knowledge deny or improperly minimize the role played by this reality. Occasionally sociologists may have given this impression, but by beginning with observation we want to emphasize that this would be a misreading of the position to be developed here. The aim will be to show how the sociological analysis of knowledge can and must proceed on the assumption that at the basis of knowledge there lies a causal interaction between the knower and reality. In pursuit of this aim we must confront certain important facts about the psychology of perception. First, though, let us see what scientists themselves say about observation.

1.1 WHAT SCIENTISTS SAY THEY OBSERVE

As the word is used by scientists, the category of 'observable' things is far broader than the list of things that can be seen, heard, tasted, smelled or touched. To 'observe' – in a laboratory context – does not generally mean simply bringing the sense organs into play. It is true that when the chemistry teacher asks the pupils to observe the 'red fumes' in the gas jar, then the process of observation called for is indeed close to that of attentive looking. Distraction, haste, inattention, etc. would all produce failure, while due care and normal eyesight would usually guarantee success, though not necessarily a uniform verbal description of what is seen. Here the word 'observation' does mean something close to the deployment of the sense organs. Usually, however, the word refers to activity of a quite different order. Thus physicists will say that there are a number of ways to observe the recoil of an atom, and cosmologists will ponder over the large-scale structures observed in the universe. Others will speak of the observed flux of neutrinos and compare it with the predicted flux. On a more modest, empirical level, budding scientists in school laboratories will observe the relationship between the temperature and pressure of a gas. None of these things can be seen in the literal sense. None can be established by inspection in the way that the presence of something plausibly answering to the description of 'red fumes' could be.

Why do scientists talk in this way, given that it seems to stretch the analogy with visual inspection beyond its breaking-point? The answer is that they are marking a contrast between what they know or take themselves to be finding out, and things that they do not yet know or want to find out. Their talk thus has a pragmatic character in that it is used to draw distinctions whose precise import varies from context to context. It is always within the framework of a particular investigation, or a particular experiment, that a scientist differentiates what can be observed from what cannot. The language of 'observation' is thus contextual or, as some sociologists would express it, 'occasioned' and 'indexical'.

The contextual character of these so called 'observations' means that they will depend on all manner of presuppositions and assumptions. As the above examples indicate, these assumptions will often be of a highly theoretical character. For example, the observation of a certain flux of neutrinos will make use of sophisticated techniques of detection, and these in turn will employ theoretical generalizations about, say, radioactivity. This is a point on which most analysts of science, whether they are philosophers or sociologists, agree. It is sometimes expressed by saying that observation in science is 'theory-laden'. A more accurate statement of the conclusion would be that observation *reports* are theory-laden. The distinction between an observation and an observation report is not usually made in ordinary laboratory talk, at least, not in any general and principled way. Nevertheless, for our purposes it is a distinction worth attending to.

For each of the so-called 'observations' referred to by scientists, i.e.

for each of their theory-laden observation reports, we can pose the question: what precisely were the things seen and heard or touched or smelled or tasted, that prompted the report? In other words, we can always ask what observations (in the narrow sense) prompted the 'observations' (in the broad sense). To pose this question we need not make any particular assumptions about the locus of what is 'really' known, nor need we be assuming any doctrines about the 'foundations' of knowledge. Our motives need be nothing more than curiosity and the plausible belief that this line of enquiry will enable us to trace some of the causal influences at work in the individual beliefs of scientists. It allows us to address the question of how particular sights and sounds come to be given the theoretical interpretation that they are. This is the question of how an observation becomes an observation report. Without differentiating the two we cannot bring the question into focus. Nevertheless it may be asked: how real is the distinction between an observation and an observation report? Could it be that when an observation receives expression in an observation report, the way it is described reacts back and influences the observation itself? Indeed, is there really a level of observation that can be differentiated from that of interpretation at all? To answer this we need to look at the relevant data from experimental psychology. Until we reach a decision on this question we cannot make the first step towards an adequate, naturalistic description of scientific knowledge.

1.2 THE EXPERIMENTAL STUDY OF PERCEPTION

At the beginning of his book *Patterns of Discovery* (1965) Norwood Russell Hanson posed the question: supposing Johannes Kepler and Tycho Brahe were to watch the sun rise, 'Do Kepler and Tycho see the same thing in the east at dawn?' (p. 5). This formulation is not meant to call into question the assumption that it is one and the same object – the sun – that they are looking at. In that sense we may suppose that they see the same thing. The object of which they have an experience, we may say, is one and the same. The point is: do they have the same experience of that object? This comes into question because, of course, Kepler and Tycho had different theories about the behaviour of the sun. For Kepler the sun was the centre of the solar system and the earth rotated around it in an annual cycle. The phenomenon of night and day was due to the additional daily rotation of the earth on its axis. The sun itself did not move. Tycho was more conservative, and adhered to a sophisticated variant of the old earth-centred picture.

While he agreed with Kepler that the familiar planets of the night sky might revolve around the sun, nevertheless he held that the sun itself revolved around the stationary earth. For Tycho, then, the sun really moved across the sky; at dawn it could be said literally that it 'rose'. For Kepler this movement was merely apparent. It was the earth that really moved, our horizon dropping away from the sun. The relative motion between the horizon and the sun was thus given two different theoretical interpretations. (For a clear account of the rival Keplerian and Tychonic systems see Kuhn's *The Copernican Revolution*, 1959, p. 202.)

Could it be that the two different theoretical understandings of what lay before them resulted in the two observers having qualitatively different experiences of the world? Did reality look different, or is the look of things stable and independent of our thoughts and interpretations? Hanson argues that Kepler and Brahe would *not* have seen the same thing. The world would look different because of the different theories. Observation, even in the narrow sense of the word, is not a neutral court of appeal between the two but, from the beginning, is shot through with interpretation: 'theories and interpretations are "there" in the seeing from the outset' (p. 10). In reaching this conclusion he cites the work of experimental psychologists, and this is why his discussion can serve to introduce this phase of our enquiry. If we use the conventional hierarchy of mental faculties, in which the thinking and believing of the cognitive faculty are 'higher' than the perception and sensation of the sensory faculty, then the question is: to what extent is the lower faculty influenced and permeated by the higher faculty? Is the lower faculty encapsulated or permeable?

Hanson uses evidence from the Gestalt psychologists to guide his analysis of the Kepler-Tycho case. Gestalt psychologists drew attention to the 'shift of aspect' that can take place when looking at certain ambiguous figures, such as the well-known Necker cube (see Figure 1.1).

This line drawing of a cube can be seen in two ways: as if from above, and as if from below. Most people looking at this figure see it one way for a while, and then experience a slightly uncanny shift to the other way. It is as if the brain is given the 'data', in the form of the lines, and has an 'hypothesis' about what it is seeing. It then 'realizes' that the data is also consistent with another hypothesis, and switches to that. Our experience in the two cases is subtly different. We seem to experience not just the lines on the page but the 'hypothesis' that the brain is currently adopting. The hypothesis is, as it were, built

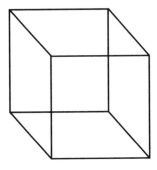

Figure 1.1 The Necker Cube

into the phenomenology. Hanson argues that the difference between two people who are looking at the Necker cube, where one sees one aspect and the other the other aspect, is like the difference between Kepler and Tycho. On this basis he denies that perception is a two-stage process, with input from the first, perceptual stage passing upwards for cognitive interpretation. Rather, our theoretical knowledge is woven into the fabric of experience itself (p. 22). In general, he concludes: 'observation of *x* is shaped by prior knowledge of *x*' (p. 19).

There are other sources of evidence pointing to this conclusion. In his *Structure of Scientific Revolutions* (1970) T.S. Kuhn quotes the work of the psychologists Bruner and Postman: 'On the Perception of Incongruity: a Paradigm' (1949). Using playing cards as stimuli they studied how accurately and with what degree of difficulty subjects were able to identify anomalous cards. Most of the stimuli were standard cards exposed for a short but carefully measured time. Occasionally a non-standard card (such as a black four of diamonds) was presented. Subjects seemed to take appreciably longer to identify these anomalies, and sometimes even seemed to see them as normal cards. Our customary expectations, attuned to the conventional format of the cards, appears to structure what we can see and do see.

Relevant data also comes from the study of optical illusions such as the famous Müller-Lyer figure. Here lines of the same length can be made to look different lengths by adding arrowhead-like additions to their ends (see Figure 1.2).

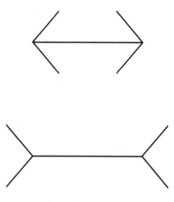

Figure 1.2 The Müller-Lyer lines

The line with the inward-pointing 'arrowheads' appears to be longer than the line with the outward-pointing ones. In his book *Eye and Brain* (1966) the psychologist Richard Gregory explains this effect in terms of the brain interpreting the 'arrowheads' as distance cues. It reads them as giving it the sort of data that we can get if we look at the corners of rooms or buildings. (Looking at the upper corner of a rectangular room where the walls and ceiling meet will typically furnish lines similar to the inward-pointing 'arrows'.) Interpreted in this way the shapes can carry the message that one of the lines 'stands out', while the other 'stands back'. Since both the lines subtend the same angle on the retina the one that is understood to be further away must be larger, so the brain makes the correction that seems called for and we experience the end product of its calculations and assumptions, and see it as larger. As Hanson said, our observation of the lines is shaped by this geometrical knowledge, only here the assumptions are inappropriate, so what we see is an illusion.

The generality and significance of data such as this has been the subject of much discussion. An example of two, divergent assessments of it is provided by the clash between Fodor and Churchland. In his *The Modularity of Mind* (1983) and his paper 'Observation Reconsidered' (1984), Fodor argues that the mind is made up of a number of special purpose systems. Its architecture, he says, is 'modular'. Thus we have systems that permit us to discriminate smell, systems that permit us to perceive spoken speech, and perhaps a special module devoted to facial recognition. In other words, we have something akin to the old system of mental faculties. One particularly important faculty is the visual system by which we transform sensory input into the picture of a world of stable, three-dimensional, coloured objects. Fodor insists that this process is largely distinct from our handling of beliefs, theories and preferences of the kind that might be entertained in a linguistic form in the course of scientific education or research work. Clearly Fodor's model is precisely the one that Hanson was attacking. Its implication is that Kepler and Tycho shared a common perceptual apparatus that will have been insulated from their scientific beliefs. In virtue of this modularity they would have seen the same thing in the east at dawn.

As evidence for this Fodor reminds us that illusions like the Müller-Lyer work even though we are familiar with the figure, and know that the lines are really of the same length. If our perception were permeated with knowledge it would be influenced by this fact, and we wouldn't continue to see the illusion once we knew it *was* an illusion. The sheer durability of effects of this kind speaks for the autonomy of the visual system and its independence from such knowledge and belief. It is important to realize that in emphasizing this point Fodor is not taking issue with the idea that our perceptual experience has the character of hypothesis formation, nor does he deny that it depends on assumptions and presuppositions. All he needs to make his point is that these 'assumptions' and 'hypotheses' are of a limited and relatively fixed kind: they cannot be drawn from anywhere in the total network of beliefs of the perceiver. So it is not the inferential character of perception that he opposes but the idea that the premises of these inferences might be furnished by, for example, the scientific theories that we may adhere to. The assumptions or theories built into our perceptions come only from a certain set that has been built into the 'module' itself, or has been acquired through its own autonomous working and development.

What about the experimental evidence that apparently tells against this view? This is largely the product of the so-called 'New Look' psychology begun in the years after the Second World War. Fodor is sceptical about its quality. As proof of the permeability of perception he doesn't believe it can withstand critical scrutiny. In particular he notes that it is difficult to know if work of this kind is really tapping the perceptual system or whether it is measuring something else entirely.

Let us take the general case that covers experiments such as Bruner and Postman's on the anomalous playing cards. Suppose an experimenter claims to have found a case in which a subject *S* is seeing what he or she expects to see. The basis of this claim is something that *S* said or did – some 'response', such as a verbal identification. How do we know that this choice of words is really a response to what is *seen*, rather than a statement or expression of what *S believed* would be there? These would seem to be quite different underlying processes, and yet in many circumstances they would generate the same verbal responses and hence the same data for the experimenter to interpret. It is Fodor's contention that the dramatic and characteristic claims of the New Look psychology about the influence of belief on perception come from experiments that don't allow us to distinguish these two possibilities: their status as evidence is therefore weak, and their usual interpretation is one-sided and contentious.

The claim here is not that these different processes could never be untangled by experiment, but merely that, as it stands, much of the evidence is open to such reinterpretation. This is a long-standing problem for certain kinds of psychological enquiry. It was a particular problem for the old introspectionist psychologists of the early years of this century, who called on their subjects to report on the contents of their conscious minds, their images, sensations and feelings. Like art students who are encouraged to paint what they see rather than what they know, the subjects of such experiments had to describe (say) their inner, subjective experience of some scene before them. But how could the psychological investigators be sure their subjects were describing the desired mental content, rather than merely reporting what they knew or believed about the scene in question? The introspectionists even had a special name for the mistake of pronouncing on the object of experience, rather than describing the experience of the object. They called it the *Kundgabe* error (cf. Humphrey, 1963, p. 120 ff). Fodor doesn't use this term, but he is in effect saying that the New Look psychology was flawed by its failure to rule out or minimize this danger. We can see at once that this problem might beset attempts to answer Hanson's question about Kepler and Tycho. If we seek from them descriptions of 'what they experience', any divergence in their reports could reflect the divergence in their theories rather than in their experience.

Such criticisms have not gone without reply and the debate continues. Nevertheless, it has proven surprisingly difficult for those who want to defend variations in actual experience to avoid the *Kundgabe* error. They must keep experience separate from the verbal rendering or labelling or theorising that is routinely prompted by that experience, and yet at crucial points of the argument that distinction is not respected; we are given evidence that could bear on the understanding of the object that is experienced, rather than the experience.

One of the most spirited defenders of the idea that scientific theories might restructure experience itself is Churchland (1988). Like Fodor's, his argument is deeply informed by respect for the relevant psychological data. Churchland advances a number of factual claims. First, he remarks that the stability of the Müller-Lyer illusion is only relative. The illusion only works because we have *learnt* that certain geometrical forms can be relied upon as distance cues. This implies that the visual system or 'module' must be permeable. Second, he argues that even the relative resistance of the Müller-Lyer illusion to interference from current knowledge or belief is hardly typical. Other optical effects can be modified by decision and choice. Churchland contributes some intriguing new examples to the discussion, but the familiar Necker cube can be used to make the point. We can, he says, learn to switch the usual aspect of such figures more or less at will, and see the cube now from above and now from below. Doesn't this again show that the visual 'module' is permeable? Third, he draws attention to the fact that we can cope with quite radical reorganizations in the relationship between our visual input and the motor system. The evidence comes from heroic experiments in which subjects have learned to live with an inverted visual world, produced by wearing special inverting lenses. The new perceptual regime can evidently be established within about a week, provided the subject makes an *active* attempt to learn to cope. Cases like this, says Churchland, 'reflect the plasticity of some very deep "assumptions" implicit in visual processing' (p. 175). The theme of activity and active use which is present in this type of relearning is an important one. Churchland does not defend the idea that any passing whim might lead to perceptual modification. Such modification will only result from the attempt to give practical application to the new perspective, and this application may also need participation in a group of like-minded fellows. Kuhn implied as much when he associated the restructuring of perception with a collective change in the practice of the scientific community, namely a change of 'paradigm'. Fourth, Churchland points out that the study of the physiological pathways in the brain suggests the existence of both ascending and descending connections, i.e. both from the 'lower', sensory faculties to the 'higher' centres, and vice versa. It would appear that, anatomically, we are so equipped that our thoughts could influence our perceptions as well as our perceptions influencing our thoughts.

Churchland writes as someone who has great faith in the capacity of science to progress and, as it does so, to transform the character of the knowing agent. Just as our visual experience is excitingly transformed by instruments such as microscopes, so perhaps the very texture of our brains will be modified by new theories, new discoveries and our ability to focus on new properties and relations in the world around us. His detailed argument ends with an affirmation. The history of science is like a 'long awakening':

> The human spirit will continue its breathtaking adventure of self-reconstruction, and its perceptual and motor capacities will continue to develop as an integral part of its self-reconstruction. (pp. 186–7)

Just as the musically educated listener, with cultivated discriminations and a trained ear, will undergo a qualitatively different experience from that of the untrained listener, so will the scientist's experiential contact with the world undergo a similar deepening and qualitative change.

Fodor is unimpressed, and gives a point-by-point reply (1988). First, the Müller-Lyer illusion again: yes, it may be learned, and that may indicate a measure of permeability of visual experience by background knowledge. But does it suggest a degree of permeability that might significantly impinge on scientific disputes? Surely not, says Fodor. Having access to knowledge about the geometrical projections of three-dimensional objects onto two-dimensional surfaces (such as the retina) represents a very limited form of permeability that will be, for the most part, scientifically neutral. Second, what about the alleged ability to manipulate illusions such as the Necker cube? We *can* do this, says Fodor, but not merely by taking up attitudes or making decisions. We have to do things like changing our fixation point deliberately to flip the visual aspect. That is, we learn how the module works, and then exploit its workings. This doesn't count against the autonomy of the system, but for it. We cannot modify what will be an ambiguous figure for us, nor what will be the effect of finding a technique for setting the interpretive machinery in motion to alter its aspect. All of this is beyond our conscious control. Third, we come to the inverting lens case. This looks like strong evidence against encapsulation. Fodor replies that, on reflection, perhaps we should have expected a breakdown of encapsulation at just this point. Doesn't, say, hand–eye coordination have to be continually recalibrated in the course of natural processes like growth? So won't there be a built-in mechanism for this? Perhaps inverting lenses are just an extreme case of an entirely natural process that evolution will have already catered for. To concede this, Fodor insists, is not to concede that we have any mechanisms that produce such adjustments in response to, for example, our believing such and such a theory. The latter process would not be expected on 'ecological' grounds and, he insists, no plausible examples have yet been given. Fourth, on the descending pathways, where Churchland cautiously felt such evidence was suggestive, Fodor is dismissive. We simply don't know what these neural structures do, and there is no direct evidence that they facilitate the 'top-down' influence of thought on perception.

In opposition to Churchland's optimistic vision of scientific progress reconstructing the human spirit, Fodor counterposes a studied pessimism. An 'endless awakening', he opines, doesn't sound much fun, and he reports a complete inability to self-reconstruct before his morning coffee. The ideology of perfectibility is met by the world-weary observation: 'Actually, theories come and theories go, and people don't really change very much; or so it seems to me' (1988, p. 198). Finally, what does Fodor make of the evidence of, say, the musical experience of the educated ear? The response epitomizes his rejection of Churchland's arguments. The phenomenon may be real enough; the musical sophisticate's experience *is* different, at least in the broad, everyday sense of 'experience'. But is the actual perception itself different, and if so why? It does not help Churchland's side of the argument to insist that the quality of perception is different if that change is just the result of, say, listening to more music. The issue is not whether perception might be educable or discriminations modifiable: it is whether it is changed by theoretical opinion or belief. Are the sophisticate's perceptions more refined – if they are – because of the musical theory that has been grasped or the acoustical doctrines adhered to? And would the experience be different if the theory changed? That is what needs to be shown and, says Fodor, it has been presumed rather than demonstrated. Surely, the main difference between the sophisticate and the non-sophisticate lies in how they label their experiences and what they say on the basis of them. Their different theories are theories about what their perceptions are perceptions *of* (p. 195). So we come back to the *Kundgabe* fallacy. Those who say that our theory of *x* alters our experience of *x* are conflating the description of an experience with a report of what it is an experience of.

1.3 A BIOLOGICAL AND METHODOLOGICAL RESPONSE

Of the positions we have just examined it may seem obvious that one is 'more favourable' to sociology than the other. For are we not asking, in effect, how far perception might be permeable by culture? And that means: how far will sociological enquiry take us? Sociologists might therefore be expected to have a predilection in favour of Hanson and Churchland, rather than Fodor. Nevertheless, our view is that, on balance, the position defended by Fodor is better supported by the evidence. The issue is by no means clear cut. For example, Fodor's appeal to the continuous recalibration of hand-eye coordination during maturation is ingenious but, as a reply to the inverting lens data, hardly compelling. Again there are cases that can be drawn from the history of science where the best explanation does seem to be that certain observers were authentically seeing what they believed *because* they believed it. For example, there is the notorious N-ray episode that took place at the turn of the century, where a number of French scientists believed they had detected a new 'ray' comparable to the then novel cathode rays, alpha rays and X-rays. The new rays were detected by the experimenter visually registering the slight change of luminous intensity of a chemically coated screen in a darkened laboratory. They were operating at the very limits of sensory detectability. Prior beliefs may well have adjusted sensory thresholds and modified the statistical process of signal detection within the noisy visual system – (see Nye, 1980). This case, however, caused immediate suspicion within the scientific community, and was uncovered by very elementary controls introduced by a sceptic. The conclusion, then, is not that scientists *never* see what they believe simply because they believe it – sometimes they seem to do just that. The conclusion is that if we intend to illuminate the central core of well-conducted experiments, and enduring scientific achievements, we would do better to accept the autonomy and stability of sense experience.

The rather specific arguments that characterize the Churchland–Fodor debate may be supplemented by an important general consideration, one that was hinted at by Fodor's reference to the 'ecology' of perception. This is the general evolutionary argument in favour of modularity: it carries survival value in a way that a perceptual system sensitive to shifting beliefs would not. Evolution surely favours a *passive* perceptual system rather than an *active* or a *creative* one. An organism that did not possess a perceptual system that could surprise and override its cognitive assumptions and expectations would hardly live to pass on its complacent and day-dreaming genes. There are, of course, limits to what such a general argument as this can achieve. Strictly it could only apply to organisms confronting something close to the normal run of everyday experience. Evolution, we might say, will not have foreclosed its options, or mapped out a response in advance, for highly novel circumstances. Nor do evolutionary considerations indicate a complete passivity on the part of our perceptual apparatus. It is biologically useful if highly probable occurrences are registered somewhat more rapidly than improbable ones. A little less information should suffice for perception in the one case than the other. Biologically, then, the best perceptual system would be a 'biased passive' one – (see Broadbent, 1973). This limited bias seems sufficient to account for the data, e.g. Bruner and Postman's anomalous playing card experiment cited by Kuhn, and the facts drawn from the history of science where occasionally scientists simply see what they expect to see. But such a biological perspective generally favours an encapsulated, largely autonomous perceptual system.

Methodological prudence also indicates that we should opt for the stability of perception. If we can show that the sociology of knowledge can make good headway under these circumstances, then we will have made a stronger case than if we had helped ourselves to apparently more favourable assumptions. Should it then happen that new psychological evidence begins to shift the balance of argument back to a greater degree of theory-ladenness of perception, then that will be a pure bonus. Exactly how we can begin to construct a sociological analysis on the assumption that perception is stable and autonomous will be shown in the next chapter. For the moment the task has been to make clear the psychological starting point of that enterprise. It will be clear that in adopting this position we are contradicting some methodological assumptions that are widely held amongst sociologists or, at least, widely attributed to them. It is sometimes said that the best way for sociologists of knowledge to proceed is *as if* the natural world, and our experience of it, played no significant role in the production of knowledge. This 'methodological idealism', as it may be called, is defended on the grounds that it will help the sociologist by letting the sociological dimension of knowledge stand out clearly. It is meant to improve the signal–noise ratio, and remove obstacles in the way of sociological enquiry. In our view methodological idealism is wrong. We will now examine a version of this position, and explain why we reject it and what we put in its place.

In his important sociological study *Changing Order* (1992 [1985]), Collins says: 'If the world must be introduced, then it should play no more role than the fire in which pictures are seen' (p. 16). He then deliberately reverses the order of common-sense thinking which treats the world as something impinging on us, rather than something projected outwards by us. Accordingly Collins recommends us to 'treat descriptive language as though it were about imaginary objects' (p. 16). Furthermore, 'all descriptive-type language should be treated as though it did not describe anything real' (p. 74). Notice the subjunctive mood. Collins is *not* saying that reality is an illusion, or that it plays no role in knowledge. He is saying that it is useful for certain purposes to proceed *as though* it were like this. Why is he saying this? He is trying to sensitize us to some very real methodological dangers, e.g. taking accepted descriptions of reality as unproblematic, and focusing the explanatory enterprise on deviations or departures from what we take for granted. There are many voices which would encourage us to adopt some version or other of this one-sided and blinkered form of curiosity – (see for example, Lakatos, 1971; Laudan, 1977). It is in an effort to stop mistakes such as this that Collins advances his methodological idealism. By trying to bracket off the effect of reality we will be helped to put all the different accounts of reality on a par with one another. Hence we will be encouraged to find them all equally problematic, and the one-sidedness of our curiosity will be overcome.

The question is: have these excellent procedural intuitions found expression in the right methodological principle? In support of Collins we may notice that experimental psychologists seem to adopt similar methods. They too sometimes try to minimize or remove the effect of external reality as an experimental strategy. For example, in studying how the brain works when we read, experimental subjects could be given written sentences, with a word left out. The aim would be to see if, and how, the brain fills in the gap. They might be given a glimpse of the words, 'In the evening he went to the'. In these circumstances some subjects report seeing the sentence 'In the evening he went to the cinema': the gap is filled by a likely candidate. Taking the report at face value, as an authentic description of perceptual experience, then it suggests that the brain subserves reading by making probabilistic adjustments to thresholds of word availability. But, whatever the correct interpretation of such results, don't we have here a version of Collins's methodology? This too looks like a revealing projection onto a blank world; a response without an external stimulus.

Certainly this experimental technique and Collins's methodology have things in common. They are both methods of subtraction. If an effect is made up of two components x and y, then if we remove factor y we can find out the x component. Thus the psychologist removes the external stimulus in order to reveal more clearly the underlying response tendencies. Similarly, Collins is telling us to imagine the input from reality taken away in order to see how much purely social processes (e.g. traditions, shared assumptions, interests, etc.) might achieve on their own. There is, however, a crucial difference. Unlike the psychologist, Collins is inviting us to perform a thought experiment, not a real experiment, and the method of subtraction does not yield determinate results when applied to thought experiments. Your conclusion about what you will be left with, when you imagine y removed from the sum of x and y, depends on what you think their proportions were in the first place. Such a procedure cannot, therefore, take us any further than we were at the outset. Methodological idealism does not give us a genuine technique for exposing the social element in knowledge: it only provides a means of expressing our prior assumptions on the matter. It has the disadvantage that it invites us to make unchecked suppositions about the scope and role of social factors, without providing any genuine controls on them. Whilst it may indeed help us distribute our curiosity evenly by correcting a one-sided perspective, this advantage is surely outweighed by these dangers.

We need an orientation that neither limits our curiosity nor requires us to make implausibly idealistic, or frankly counterfactual, assumptions. Furthermore we require it to be consistent with what psychologists can tell us about those causal interactions between knower and known that we call 'observation' or 'perception'. Such an orientation can be provided if we start the analysis of knowledge by attending carefully to the aspects of the world that are registered by the knower, while remembering that how these perceptions are acted upon and linguistically described, categorized, labelled and classified remains to be determined, and remains to be explained. In other words, we must fasten firmly onto the fact that the cultural response is underdetermined by the psychological stimuli. For this reason even the most mundane instances of observation are capable of providing material for sociological analysis. A person doing something simple, such as responding to a red object, would suffice, e.g. a car driver at the traffic lights, or a chemist with a piece of litmus paper.

Think of the observer's eyes in such a case as instruments, i.e.

detecting instruments. Now, in practice it is clear that instruments (e.g. microscopes, photo cells, Geiger counters, etc.) are always used with an awareness of their fallibility and vulnerability. The deliverances of any particular instrument can never be assumed to be infallible. (Should they be so treated, on some special occasion, then that is a status assigned to them, and it could always be withdrawn.) The same applies to any individual observer. The fact that there are arguments over something as apparently simple as seeing a red object should be enough to establish this. Observers, like instruments generally, only yield up reliable knowledge when they are working properly. If we ask where the criterion of proper functioning comes from, we find that it lies in the coherence of the individual observer or instrument with other relevantly similar ones. However self-evident or subjectively convincing an observation seems to the observer, it is only counted as genuine or veridical if it coheres with the generality of other readings. Individual and internal criteria don't outweigh collective ones. The standard implicit in treating an observation as genuine or reliable means that there is a social criterion to be satisfied before something can even *count* as a genuine observation for its users.

It is no fundamental objection against this view to point out that we can easily imagine the possibility of one observer or one instrument working correctly even if it stands in isolated opposition to the deliverances of all the others. The objection would only succeed against a very naive version of the doctrine that standards of acceptability are shared and consensual. The process of consensus formation doesn't take the form of a simple majority vote at some specific point in time. Any such snap shot could indeed yield up results in which the isolated deviant or critic would later turn out to be right. The ultimate correctness of the consensual view shows itself, however, when we ask: what would this 'turning out to be right' amount to? The answer is that it would be a new consensus. Although we may *seem* to operate with a non-social, non-consensual idea of the proper working of a detecting instrument or observer, really we are simply making a longer term, higher order reference to the relevant collective.

We do not therefore have to wait until we are in the realm of high theory or daring speculation before social processes come into play. Nor do we have to assume that observation is strongly influenced by expectations, or that the causal influence upon us of material objects should be wished away. Even as a methodological exercise there is no need to play down the role of external stimulation, because the goal is simply that of seeing how the individual's commerce with reality is taken up into the patterns of interaction between people – all of whom are in their turn interacting with reality. Until the individual response is so taken up, we will not be in a position to discern those features of knowledge that are familiar to us, and characteristic of it. The process just referred to – in which individual responses are taken up into patterns of social interaction – is sometimes known by another name: it is called 'interpretation', and it is to this that we now turn.

CHAPTER 2

Interpretation

2.1 INTRODUCTION

In this chapter our aim is to take a single experiment and subject it to sociological analysis. For this purpose we have chosen Millikan's famous oil-drop experiment. In this experiment, first performed in 1910, the Nobel prize-winning American physicist R.A. Millikan (1868–1953) measured the fundamental unit of electric charge, that is, the charge on the electron itself. There are four reasons for choosing this case. First, the experiment is a classic. It depended on simple physical principles, and yet it gave the accepted value of one of the most important physical constants. But Millikan didn't merely measure a property of the electron: his experiment also helped consolidate belief in its physical reality, and confirmed its status as the atomic unit of electricity. Second, the experiment was for a while the subject of a sharp controversy. No sooner had Millikan begun to publish his results than they were challenged by the Viennese physicist Felix Ehrenhaft (1879–1952). The Millikan–Ehrenhaft debate helps throw into relief the character of Millikan's reasoning and the methods he used in the interpretation of his results. Third, these issues have been subject to a detailed and sophisticated analysis by the historian and physicist Gerald Holton who has had access to Millikan's laboratory notebooks (Holton, 1978). He has been able to give us a glimpse of the moment-by-moment judgements and decisions that lay behind Millikan's published findings. Fourth, Holton's paper has itself been subject to a degree of criticism by Franklin (1986). This will help to highlight a number of important methodological themes and problems characteristic of work in this area. In what follows we will depend heavily (and gratefully) on Holton's important paper, but we will be extracting from it conclusions that are somewhat different in their emphasis from Holton's own. His main concern was to exhibit the 'themata' in scientific work, that is, its theoretical and philosophical commitments. These are certainly of central importance, but our concern will be to ask what these commitments can reveal about social processes.

In the next section we will describe the experiment and then go on to discuss its result. It is worth noting that even innocent references such as this to 'the experiment' already use social categories. There was, of course, no simple, unitary episode which constituted 'the experiment', nor was there a single outcome that constituted 'the result'. There is no defensible sense in which the complex sequence of actual events that are brought under these headings were solely Millikan's doing, or his sole responsibility. This does not mean that the use of the definite article is wrong; rather, it reminds us that 'the' experiment and 'the' result are *institutions*. All manner of real-life complexities have been filtered out as Millikan's efforts increasingly assumed the status of an accepted finding, and a shared reference point for the community of physicists. Our aim will be to exhibit some of the complexities and contingencies of the process, and to explain their sociological significance.

2.2 MILLIKAN'S OIL DROP EXPERIMENT AND THE CHARGE ON THE ELECTRON
The experiment
Imagine a tiny droplet of oil falling through the air under gravity. After a brief period of acceleration, the downward pull of gravity will be balanced by its buoyancy and the retarding force of friction. It will then fall at constant speed. There is a well-known law, called Stokes' Law, which relates the force of friction to the viscosity of the air, the radius of the drop and the speed of fall. This means that the physicist can begin to write down equations linking together some of the features of this situation. Now suppose that the physicist arranges for the droplet's fall to take place between two horizontal metal plates that can be electrically charged. If the droplet itself possessed or acquired a charge it would interact with the plates, being attracted or repelled according to their polarity. If the drop were negatively charged, and the top plate were positive, its fall could be arrested or even reversed. The attraction upwards would offset the weight of the drop. Given the right charges on the plate the droplet could be held in balance. Since the charge on the plate could be measured, this would enable the physicist to calculate the charge on the droplet: it is that charge which, in conjunction with that on the plate, creates an upward force which exactly equals the downward force of the weight. But how can we know the weight of such a tiny droplet? That is where the previous equations come in. The Stokes' Law equation allows us to calculate the radius of the drop and, assuming it is spherical and that we know the density of oil and the viscosity of air, we

can easily calculate its weight. Having got this, the charge on the drop can be found.

How might the droplets acquire their charge? The answer is that the droplets would have to be produced by some form of spray or atomizer, and the friction of passing through the nozzle of the spray is sufficient to charge them. In fact Millikan went further than this, and introduced another kind of electrical charge on the droplets. Using a radioactive source he ionized the air through which the droplets fell. Collision between a droplet and an ionized (i.e. charged) molecule of air meant that the droplet could itself acquire one or more electrons. If this happened to a droplet held stable between the charged plates it would suddenly upset the equilibrium, and the drop would be pulled upwards. In principle, then, the physicist now has two movements to observe and time: that of a falling droplet and that of a rising droplet. He therefore has two lines of approach with which to calculate the charge that happens to be on the droplet during these journeys. We can call the charge acquired through friction qf, and the charge acquired by ionization qi. In general they will not be the same, but if all of these charges are got by adding and subtracting identically charged electrons, their differences should be some integral multiple of the charge of the electron. If the experimenter surveys all the various charges that he finds on many droplets, then the chances are that the smallest charge will be that of the electron itself. He should then find that all the other charges are some whole number multiples of this. Suppose we call the smallest observed charge e; then we should find that qf is equal to (say) ne, and qi equals (say) me, where n and m are 1, 2, 3, 4 etc. To repeat: the crucial features of the results will be (i) the size of the smallest charge found, and (ii) the fact that all the other charges should be definite multiples of this, with no intermediate or fractional values.

So far only the bare, abstract principles of the experiment have been described. In practice all sorts of difficulties need to be overcome. For example, the calculation depends on Stokes' Law, but this law only applies strictly under idealized and unrealizable conditions. Again, if the drop were to evaporate during the course of the experiment the calculation would be made inaccurate. An unrecognized fluctuation in the voltage on the plates would also have a marked effect on the calculation. Clearly the whole apparatus must be carefully shielded from draughts, and care must be taken to avoid convection currents between the plates. If the droplet is being carried by moving currents of air the theoretical analysis will be inapplicable. But even if gross movements of the air can be removed, there is also the danger that the molecular impacts that cause so-called 'Brownian Motion' will make themselves felt. The longer the fall the greater the probability of some such interference; but, conversely, the shorter the fall of the droplet the more difficult it would be for an observer to time with a stop-watch.

As a scrupulous and seasoned experimenter Millikan was well aware of these problems, and continually sought to remove sources of error. Thus he shifted from an initial study of water droplets to oil droplets, precisely to cut down evaporation. He was also very careful to shield the apparatus, to keep the voltages steady, and to keep dust out of it. (Dust was, in fact, a considerable problem, and great efforts were made to keep the apparatus free of it because it significantly affected the readings.) Having taken all due precautions he was confident that he knew precisely what was going on within the sheltered confines of his apparatus. Whereas previous researchers had concerned themselves with the study of clouds of droplets, he could now study individual particles and spot the precise moment at which an electron attached itself to a droplet. As he put it: 'The whole apparatus . . . represented a device for catching and essentially seeing an individual electron riding on a drop of oil' (Holton, 1978, p. 37).

Millikan wasn't saying that the observer in his experiment can actually see electrons. He said that 'essentially' he could see one. The word 'essentially' has the implication that what we *actually* see (e.g. an oil drop suddenly moving upwards) is *really* the catching of an electron. Elsewhere Millikan said: 'He who has seen that experiment and hundreds of investigators have observed it, [has] in effect SEEN the electron' (Holton, 1978, p. 37). Again, despite the capitalization of the word 'seen' the claim is really the qualified one that the electron has 'in effect' been seen. This is meant as an identification of the object, not a description of its appearance. It is an interpretation not an observation. The most that can be (literally) observed is the moment at which something happens, and that something is said to be – is interpreted as – the capture of an electron.

The result
As a result of his measurements Millikan reported that the charge on the electron was approximately 4.7×10^{-10} esu. This was the smallest charge he detected and the basic unit or building block out of which all the other charges seemed to be built. In fact his findings varied slightly over the years as he sought to refine the experiment. In 1909 it was 4.69,

in 1910 it was 4.9016, in 1911 4.891 and in 1913 4.774 × 10^{-10} esu; but it was the figure of 4.7 that was taken up by Bohr and used in his famous 1913 work 'On the Constitution of Atoms and Molecules', in which he developed his picture of the atom as a miniature solar system with negatively charged electrons orbiting around a positively charged nucleus (1913). (For an account of the genesis of the Bohr atom showing Bohr's use of Millikan's work, see Heilbron and Kuhn, 1969.)

How exactly did Millikan arrive at these figures? Part of the answer, of course, is that he looked with great attention down a viewing telescope focused on the individual oil droplets, pressed a stop-watch and jotted down times and voltages and the like. These readings were entered into his laboratory notebook and then substituted in various formulae in the way described above. All this goes without saying. However, when we ask 'how did he reach these conclusions?' we mean: what happened immediately after this? Was there a smooth, automatic, unproblematic path joining the readings entered into the laboratory notebook and the data in the published papers that were used by Rutherford, Bohr and the rest of the scientific community? As Holton makes clear, the answer to this question is negative. The route from the notebook to the published paper was complex and interesting. The reason is that some of the experimental runs were counted as good ones, while others were counted as bad and discarded. An element of selectivity and judgement enters into the story. Until this is fully understood we will not have a picture of Millikan's scientific procedure that is remotely realistic. The reason why it is vital to achieve such an understanding is because, if these selections and judgements had been different, then his final result would have been different. His conclusions were sensitive to how he treated his data.

Here are some examples of these judgmental processes that Holton located in the notebook Millikan used to produce his 1913 paper in the *Physical Review* (1913). In the published paper there is a table, labelled Table XV, that deals with the so-called 'drop no. 41'. In his laboratory notes Millikan recorded that he observed this drop during an experimental run beginning at 4.15 pm on Friday, 15 March 1912. He first observed it fall under gravity (taking about 24 seconds), and then watched it rise under the influence of the electric field, again carefully timing its journey. Then he let it fall again and repeated the process, only the next upward journey took less time because of extra charge that the drop must have acquired by contact with a gas ion during the previous fall. He then went through the calculating procedure, sketched above, of using the times of the successive rises and falls to compute quantities that were proportional to the charges. By guessing whether he was dealing with 1, 2, 3 or more units of charge, he was able to estimate the unit of charge *f* as derived from friction (call it *ef*), and the unit as derived from ionization (call it *ei*). For drop 41 the whole process worked extremely well, and the desired agreement of *ef* and *ei* was very close. In other words, his two different ways of calculating the unit of charge had converged on one another, giving the satisfying impression that he was, as it were, seeing the same object from different angles. He wrote at the corner of the page, 'Beauty. Publish this surely, beautiful'.

The next run, on another drop, was not so good. This drop appeared to have been heavier (as indicated by the shorter time of fall), and it seems that it didn't prove possible to keep track of it for as many descents and ascents as the previous drop. Here Millikan's quick numerical check scribbled in the notebook revealed that *ei* and *ef* were not as close as before. 'Error high, will not use' notes Millikan, and this run was passed over and didn't find its way into the printed paper. There are, says Holton, many other observations like this that never saw the light of day. He gives a list of Millikan's comments that accompanied such rejections, e.g. 'Very low. Something wrong''; 'Possibly a double drop'; 'This seems to show clearly that the field is not exactly uniform . . .'; 'Something the matter . . .'; 'Agreement poor. Will not work out'; 'This drop flickered as though asymmetrical'; 'Can't get differences'; 'Brownian comes in . . .'; 'Something wrong with this'.

To appreciate the effect of these judgements it is important to realize that, if we allow some latitude for experimental error, *any* list of numbers representing a range of observed values for the charges on the droplets can be represented as integral multiples of a supposed unit value, *provided the unit is made small enough*. More specifically, as Holton points out, in the case of the rejected droplet observed in the second run on 12 March 1912, it is possible to get the desired concordance between *ef* and *ei* if we assume that charges accumulate in units about one tenth the size that Millikan attributed to the electron (Holton, p. 68). The effect of Millikan's selection was thus to rule out certain possible, though very small, values for the unit of electric charge and to sustain another, larger value. To express the point in another way: Millikan was using a prior belief that the unit of charge was in the region of 4.7×10^{-10} esu to evaluate his own procedure. He was therefore begging the question of whether or not there might be sub-electrons with a charge of, say, one tenth that value.

What was it that made the discordant values that he calculated for *ef* and *ei* a sign of error in his measurements, rather than a sign of error in the assumptions that he brought to the calculation? Why didn't he use these observations to reject his assumption about the approximate size of the unit he was seeking to measure? Part of the reason was that previous attempts, by himself and others, to estimate the size of the unit had produced answers in this region, though there was considerable variation between them. Another possible reason is that the experimental runs that were discarded were suspect *on quite independent grounds.* Having observed hundreds of droplets, Millikan would naturally have become sensitive to the subtly different appearances of droplets as they descended and ascended. He would have found that certain oddities of appearance in such journeys could be tracked back to, say, fluctuating batteries, or the presence of dust, or a malfunctioning spray. Nevertheless the question must be pressed further. Were *all* of the cases in which Millikan discarded readings ones where he had grounds for suspecting the apparatus, and grounds that were independent of the readings it was producing? The answer given by Holton is negative: sometimes Millikan deduced that something must have gone wrong simply because the results were not as they should be. Here Millikan used what Holton (p. 54) calls 'plausibility arguments'. This refers to the fact that there is a list of ways in which the apparatus is known to have gone wrong, and a list of ways in which it is known that it *could* go wrong, and an unexpected and unwelcome result is attributed to one of these. Thus Millikan noted 'Something wrong', but what that something is was never specified or tracked down. It is left as a bare 'something' whose existence is inferred from the 'wrongness'.

Ehrenhaft pointed out that there is a logical circle here, and he suggested that this question-begging procedure is unscientific (Ehrenhaft, 1941, pp. 406 and 443). We might even go further than Ehrenhaft, and wonder whether Millikan was simply 'cheating'. Now, it certainly wasn't Holton's intention to expose cheating, and he doesn't take himself to have done this. Rather, he takes Millikan's procedure to be no more than a bit of inferential risk-taking of a kind that is inseparable from scientific theorizing (Holton, 1978, p. 54). As far as it goes this is surely correct. Any attempt to interpret the world theoretically must, at some stage, impose its categories and meanings, and at this point, inevitably, it will be both risky and dogmatic. Logically the dogmatism will betray itself in question-begging argument. Ehrenhaft's logic was astute, but his complaint should have been directed not against the general scientific character of Millikan's procedure, but against the particular kind of science he was engaged in – and this, indeed, was where the main thrust of his attack lay. More of that later.

The interpretive procedure that Ehrenhaft and Holton bring to light in Millikan's work is a special case of a process that informs *all* activities of sense-making, from the most scrupulous to the most unscrupulous. It goes by different names: some writers call it the 'hermeneutic circle', others follow Karl Mannheim and call it the 'documentary method'. Mannheim asked us to reflect on the predicament of someone trying to understand some fragments of a document. If they knew the import of the whole document, then they could make sense of the fragments. Unfortunately, the only access they have to the whole is by extrapolating from the meaning of the fragments. The whole must be understood on the basis of the parts, and the parts must be understood on the basis of the whole. There is no way to break out of this circle except by making a conjecture and showing how, on the basis of the conjecture, a coherence can be exhibited (Mannheim, 1952; Garfinkel, 1967). We can understand Millikan's work by seeing him in an analogous hermeneutic predicament, and as adopting an analogous method. His experimental data are his fragments. The whole document is the unknown reality that underlies and produces them. His guiding theory is the meaning he imputes to the document, a theory that guides his response to the evidential fragments, deciding which are reliable, which have undergone corruption or alteration or decay or misunderstanding.

2.3 A SOCIOLOGICAL READING

What is 'social' about the episode that has just been described? The answer is that the repertoire of interpretive processes used by Millikan depended on a local cultural tradition. There was a tradition for interpreting the fragments of evidence his experiment produced. He did not have to invent for himself a sense of the underlying meaning of what he found. He could draw upon the tradition, and his success in doing so helped in turn to consolidate it. Stated in this way the point seems a very elementary one and, on some level, might seem easy to accept. The difficulty lies in realizing that it cuts deep and has profound implications. There are many ways in which the insights that can be afforded by this fact might be evaded or trivialized. For that reason we will first state our case directly and then, in the next section, address a range of objections.

Duhem long ago gave us the literary device for dramatizing the contribution of the local cultural tradition to the understanding of an experiment. He asked us to imagine what a person ignorant of science would make of an array of instruments set out in a physics laboratory (1914). His point, of course, was that without the requisite course in physics they are just so many pieces of brass and wood and glass. This clearly applies to Millikan's apparatus. Even if a stranger were familiar with oil and air, droplets and batteries, he would still be a long way from making the same sense of it as Millikan did. The gap between observation and interpretation is clear: both may observe the same thing, but the interpretation involves bringing to bear the resources of a tradition. From the standpoint of the individual the interpretive traditions of science are largely inherited from others, shared with others, validated by others and sustained in the course of interacting with others. This is not to say that people are mechanical and unimaginative in the way they deploy the theories they inherit. New applications of old ideas will always involve more or less modification, but this ubiquitous creativity must be one that others can go along with and find both acceptable and useful.

We can refine this argument by emphasizing that in science we are not dealing with a single, static tradition, but with local variants of an ever-changing tradition. Millikan used his experiments to support conclusions about the atomic nature of electricity, and the independent existence of electrons as the elementary units of electricity. Strictly speaking, however, the locus of the effects studied by Millikan was the interaction of electricity with matter (Holton, 1978, p. 132). Millikan's apparatus – as even the minimally physically informed observer can see – merely permitted electricity to interact with drops of oil. It therefore dealt with the *interface* between electricity and gross matter. To make the move from this interface to conclusions about the nature of electricity *as such* involves a special inference. What is more, this inference is one that cannot be sanctioned by the general principles of deductive reasoning nor by common sense. Not even the most unsophisticated person is inclined to think that because he draws water from a well by the bucketful, that means that water exists naturally in bucket-sized units. Water is a continuous fluid, and it is we who divide it when we scoop it up. The man–water interface, as we might call it, doesn't sanction inferences to the nature of water as such. Why then should we be so impressed by Millikan when he makes the corresponding move from the oil–electricity interface to the nature of electricity as such? Clearly a very particular, local pattern of trust has been built up. It is this trust that provides the inferential bridge, and our suggestion is that it is the local culture of physics that makes the move plausible to us. The ultimate cause of our acceptance is that a sufficient number of trusted authorities can be found who are prepared to let such an inference pass muster, or even encourage it, even though in other circumstances (e.g. standing round the village well) they would laugh at it.

For many years prior to this construction being put on the oil-drop experiment, the physics community had been aware of other phenomena at the interface of electricity and matter that likewise tempted them to try out an atomistic interpretation. As Holton points out, in his *Electricity and Magnetism* (1873) James Clerk Maxwell described the well-known phenomenon of electrolysis, and noted that it finds a ready interpretation if we think of a flow of 'molecules' of electricity. Nevertheless, Maxwell didn't take such talk wholly seriously. He observed that – useful as it was – it was out of tune with the rest of the book which developed a picture of electricity as a continuous field governed by a set of partial differential equations. The highly successful character of the field theory (it explained the behaviour of light, and predicted radio waves), suggested that the phenomenon of electrolysis would eventually find another and better interpretation that would bring it into harmony with other electromagnetic phenomena. Millikan could have adopted Maxwell's line, and adopted a cautious and conservative response to the inference that suggested itself. (Although he had reservations about it, Kelvin had already hinted at the kind of physical model that might illuminate the way a continuous electric fluid could coagulate when it comes into contact with matter. Think, he said, of liquid mercury on a table-top, or water dropped on a red-hot iron plate (Kelvin, 1897, p. 84).) But Millikan didn't hold back, so something must have changed since Maxwell's time. What had changed, of course, was the sense of where the most exciting and fruitful lines of research were to be found. That is why it is necessary to locate inferences in the context of a *local* interpretive tradition.

Millikan's treatment of his results allows us to make a further observation about how the local resources of the culture are employed. So far the point has been that for a given body of experimental results (R) there will be more than one theory (T) that could explain them. If we use an arrow to symbolize 'implies' or 'explains', then we could have $T_1 \rightarrow R$ and $T_2 \rightarrow R$, and the claim is that the preferred theory is selected because it is salient in the local culture. More needs to be said about the causes of this salience, but here we want to emphasize that the local culture also enters into the results to be explained. It enters into the R to be explained, as well as the T that does the explaining. This was demonstrated by the element of circularity that was detectable in Millikan's selection of good, publishable experimental results and his filtering out of suspect, unpublishable ones. The preferred theory for interpreting the results is also the preferred theory for identifying what really needs explaining. It isn't the only resource for this, but it is one such. In our terms, the local theoretical tradition enters into the identification of genuine facts, or genuine phenomena, as distinct from mere artefacts. It appears that there is a local factual tradition, as well as a local theoretical tradition. The notion of a local scientific culture must therefore embrace both fact and theory. This is not because we are so gullible that we see what we believe, but because we cannot evade the responsibility of interpreting and understanding what we see. Millikan was a careful scientist reading his results off his apparatus; what has been said shows what is involved in this process. The mind of the individual scientist is the point of contact between our physical environment and our social environment. Interpretation is where nature and culture come together.

2.4 OBJECTIONS AND REPLIES

The sociological reading of the Millikan experiment has, so far, depended on a single concept. We have shown how the interpretation of the experiment has depended, at both the theoretical and factual levels, on the appeal to a local interpretive tradition. No one would deny that traditions are social phenomena. Nevertheless it might be objected that this reading of the experiment is superficial. The argument could go like this: just because physicists have developed a tradition of interpretation doesn't mean that we get to the bottom of the matter by noting this fact, and giving it a sociological label. Shouldn't we be asking why the tradition has developed, or why it survives? And that – the objection continues – surely leads us straight to non-sociological considerations such as the *rationality* of the tradition, e.g. its predictive success, problem-solving power, its capacity to withstand tests and criticism and to avoid falsification, and so on through the whole gamut of intellectual virtues and good reasons.

This argument looks more powerful than it really is. Its weakness is that it trades on a false dichotomy between the rational and the social. The trick being played on us is to make us think that, when we turn our minds towards the intellectual virtues of a piece of scientific work, we are looking at something wholly different from its sociological properties. It makes us forget what has already been demonstrated during the detailed examination of the Millikan experiment. The reason for selecting such a classic piece of science was precisely that it embodies the intellectual virtues, and hence shows how they are integral to its social properties. Take, as an example, the predictive success of the electron conception as that virtue appears in Millikan's work. Anyone pressing the objection sketched above must be forgetting that we have already examined the ways in which Millikan confronted breakdowns in predictive success. We have seen how he managed to turn potential defeat into actual triumph. He cleverly used the resources of the interpretive tradition in order to find reasons for discounting problematic runs. Judging from Holton's analysis of the laboratory notebooks this even, on occasion, reached the point where the predictive success of the theory became the criterion for deciding which observations would be used to test that success. So it cannot be said that the analysis so far has avoided addressing this particular, rational property of Millikan's work.

At this point the objector might draw attention to further facts about Millikan's work and its reception, and offer these as the non-social determinants of its intellectual value, and hence as a rival to the sociological reading. For example, perhaps we should give a prominent role to the fact that (a) sometimes Millikan's expectation worked in a completely simple and straightforward way, as in the case of drop no. 41; (b) other people have been able to take up Millikan's work and use it theoretically to explain other phenomena, e.g. as when Bohr used Millikan's value for e to improve his predictions of the absorption spectra of various elements; and (c) other people have been able to repeat Millikan's experiment and get the same answer. How do these facts find their explanation if we don't conclude that Millikan simply got something right about how nature works?

Let us take each of the above points in turn. First, it is true that sometimes Millikan's expectations (e.g. that $ef = ei$) worked well, and the frequency with which this happened is bound to have contributed to the credibility of the electron theory. Is this fact a matter of sociology? Is the frequency with which ef came satisfactorily close to ei a social fact? The criterion of being 'close enough' may be deemed conventional, but the frequency with which the readings fell within these limits surely isn't. What, then, is it, and how does it stand with respect to the social reading? The answer is that, to the best of our current historical knowledge of this episode, it is to be accepted as a brute contingency. This is just what happened when Millikan went through certain motions with his apparatus. It is just one more historical fact along with the fact, say, that we happen to possess Millikan's notebooks for the 1913 experiment, but not for the 1910 one. The important point to realize however is that, just because this isn't a specifically 'social fact', that doesn't automatically make it into a 'rational fact', i.e. a fact that somehow finds its explanation in a rival standpoint, or a fact that demands a non-sociological standpoint to make it intelligible. It stands neutrally between sociological and more traditional philosophical accounts as simply one of the ways that events happened to unfold, when fortune favoured Millikan.

Is it really true that those who oppose the sociological analysis don't themselves have more to say about the successful aspects of Millikan's approach? Can't they – and shouldn't we – just say that Millikan's procedures worked because he was right? Millikan's own theory surely provides the explanation of why it worked. This is undoubtedly how Millikan saw things, and it is a very natural attitude to adopt. (Indeed, it is *the* natural attitude.) But, for an analyst who is not a physicist, and whose aim is to stand back and illuminate the workings of science, such an attitude is hardly appropriate. It would be to adopt the role of the physicist, rather than to stand back and scrutinize that role. For the scientist the world is the object of study; for the sociologist it is the scientist-studying-the-world that is the object. For these reasons alone we should avoid the inference to the rightness of Millikan's theory from the fact that it works, followed by the explanation of the fact that it works that is provided by reference to its truth. Instead of making these moves we must study how others make them, and bring to light their character.

If we make such inferences a topic, rather than a resource, the first thing we notice is that they can't be counted as deductively valid. There is no valid deductive pathway from 'theory T works' to 'theory T is true'. The reason is simple: false theories can make true predictions; false premises can yield true conclusions. Of course a scientist might choose to believe a theory because it makes some correct predictions, but that is a different matter. Historically we know that sometimes scientists take this non-deductive step; and sometimes they don't. What is needed is an explanation of why conviction sometimes goes one way, and sometimes another. As far as this line of enquiry goes, the routine, unproblematic successes that theories sometimes enjoy (i.e. successes that are not sustained by selective or interpretive work), belong to the realm of pure contingency. They are not, for the detached analyst, explained by the truth of the theory. They are not explained by anything. They are one factor among others that go towards explaining why the theory might be counted as true. But they cannot possibly be the only factor. This is proven by the fact that Maxwell was well aware that the 'molecules of electricity' picture worked for some purposes, but he still did not believe it. The sociologist's job is to seek out some of the further contingencies that illuminate the imputation of truth to a theory.

It may be clear that we can't infer truth from predictive success, but what about probability? Surely the fact that a theory works gives it a definite probability of being true, where the more it works, the more likely it is? Since there are a number of rigorous mathematical models of confirmation procedures, doesn't this show that, after all, the core of the inferential processes in science rests on rational and non-social foundations? Again, a negative answer must be given. There are many different special forms that the argument can adopt (because of the many different forms of confirmation theory); but the basic point is simple. Even if (as is plausible) the contingent course of unproblematic observation serves to adjust the belief thresholds, or degrees of belief, in our heads, the process is dependent on the interpretive traditions that have already been identified. The Maxwell example will serve again. No doubt the contingent successes of the molecular view of electricity gave it an increased probability within his system of personal beliefs. Nevertheless, it still couldn't compete with the probability given to the field theory. To explain such cases in probabilistic terms, it can be assumed that the final probability of a theory depends not merely on its capacity to explain, but also on its *prior* or *initial* probability. In other words, a theory that you find antecedently very implausible will have to be very successful to compete with one that initially seems very plausible. The question is: what sets these prior probabilities? The only plausible historical answer is to refer to the consensus of the scientific community, i.e. the local cultural tradition.

But even if we accept that the initial probability distribution has the character of an institution, won't the rational adjustment of belief in the light of evidence eventually push belief in the direction of truth? Even if scientists *begin* under the influence of different interpretive traditions (e.g. because they come from different schools of thought, rival laboratories, etc.), won't the empirical evidence gradually swamp their initial prejudices and lead to a convergence of opinion? The historical record, showing revolutionary changes in scientific opinion, may not seem to support such convergence, but it is an attractive idea, and it is always possible to gloss a sequence of historical events so that it looks progressive. There are even theorems within the probability calculus that seem to give these strong intuitions some real substance. It transpires, however, that the conditions under which the theorems hold are very unrealistic. They are far removed from the reality of scientific practice. In particular they require that the scientific evidence be sampled in a *random* manner, and that different pieces of evidence be probabilistically *independent*. Stated simply, this means that the conditions of convergence would only hold if scientific investigation were like dipping our hand into an urn, and blindly drawing out samples of coloured marbles, replacing them, and drawing further samples. It rules out all the deliberation and design that goes into the conduct of scientific investigation that makes its experiments non-random and deliberately interdependent. The distance between real-life scientific work and the statistician's caricature is a measure of our inability to guarantee convergence. There is, then, no way that the cultural presuppositions and interpretive traditions which inform experiment can be transcended (cf. Hesse, 1974, pp. 115–19).

Nevertheless, it may seem that accepting the fact that sometimes our theories work unproblematically must weaken the case for the sociology of knowledge. In our view there is no retreat involved here, but the matter certainly calls for careful handling. No plausible sociology of knowledge could deny a role for such basic, material and causal factors in the process of belief formation: sometimes theories work, and we are impressed by this. To deny this would be to adopt a form of *idealism* in which the world is understood as an emanation of our beliefs, rather than as a cause of them. That orientation has already been rejected in Chapter 1 in the discussion of observation. Acknowledging the brute contingency of the outcome of experimental procedures, and the occasional success of prediction, is no more than an extension of what has been said about observation. Not merely sensory input, but whole cycles of (externally directed) actions on the environment followed by (inwardly directed) feedback from it, act as partial causes of belief. We must take for granted that humans, like other animals, can sometimes establish reasonably reliable and tolerably non-frustrating routines for interacting with the environment. Just as birds can build nests and beavers construct dams, so humans can grow crops, prepare food, make tools, implements and useful artefacts like steam engines. The simplest examples will serve to make the point, so consider abilities like weaving cloth and making pottery. Obviously, people are here interacting with their material environment: they are not just operating in the sphere of thought, forming idle opinions or dreaming dreams. Nevertheless these are socially organized and socially structured activities in a deep and interesting sense. There are many ways of doing these things, and no unique criterion for deciding how well they have been done. They are done for many different purposes and in many different styles, and with many different understandings of how and why they have gone wrong or fallen short (because go wrong and fall short they certainly will). There are as many ways of apportioning blame for shortcomings as there are ways of diagnosing problems and suggesting improvements. Acknowledged skills in performing these tasks will be socially distributed, and the relevant expertise can only be transmitted and sustained given the requisite structure of authority. No sociologist should deny the non-social, material and physiological basis of such activities; but conversely, no account that omits the social dimension can be plausible or complete, or do justice to the historical facts.

Millikan, then, constructed a workable device (like the potter's wheel and kiln, or the weaver's loom). It gives us a reasonably reliable way of acting on the world. Little wonder that others can take it over and use it, or that its productions can act as a reference point for a developing system of theoretical understanding. But, once again, we must not treat the contingent, material core of these interactions with the world as the whole story or as giving us the essence of the matter. What is picked up and used by others as a basis for further work and theoretical calculation is not the totality of material contingencies associated with the experiment, but a selected, filtered and simplified version of them. To show this, and to show why it is important, let us look at the Millikan–Ehrenhaft debate.

2.5 THE MILLIKAN–EHRENHAFT CONTROVERSY

Shortly before Millikan produced the first version of his experiment on liquid drops (a version in which he used water rather than oil), Felix Ehrenhaft had already begun publishing his findings on the charge on tiny particles of metal (red selenium) and wax. Initially his purpose was exactly the same as Millikan's: to measure the unit of electric charge. Ehrenhaft's preference for solid particles rather than liquid drops may have come partly from his previous, prizewinning expertise in working with colloidal suspensions of such particles, but it also enabled him to overcome problems connected with evaporation. His apparatus, though not identical with Millikan's, exploited similar physical and investigatory principles, namely, timing the motion of charged particles as they move through a viscous gas under the influence of an electric field and gravity. The value that Ehrenhaft arrived at in 1909 was $e = 4.6 \times 10^{-10}$ esu, a result welcomed and cited by Rutherford. Soon, however, Ehrenhaft was reporting charges of $e/3$ and $e/5$, and later even smaller basic units were detected. Instead of picking up charges that clustered neatly around integral multiples of 4.6 or 4.7×10^{-10} esu he was finding a more even spread of possibilities. That, of course, meant that any unit charge, if it existed, must get smaller and smaller to accommodate the findings. It appeared that the apparatus must be picking up sub-electrons, but the more he pursued the work, the smaller the sub-electrons became, and the more he yielded to the suspicion that both electrons and sub-electrons might be an illusion. The correct stance for the physicist, he concluded, was to refrain from postulating any such invisible, theoretical entities. All that mattered were the law-like regularities that could be detected empirically. In sharp contrast to Millikan, who was a robust and confident realist about atoms and electrons, Ehrenhaft became a phenomenalist and empiricist along the lines of the influential Viennese philosopher Ernst Mach (1838–1916). (For an account of Mach see Blackmore, 1972.)

How did Ehrenhaft come to get results so different from Millikan's? Was there something wrong with his apparatus, or the way he analysed the events taking place inside it? Not surprisingly, Millikan had his suspicions about the suitability of Ehrenhaft's apparatus for capturing the effects under study. For example, he wondered if the particles Ehrenhaft used were truly spherical, because if they weren't, then certain assumptions used in calculating their motion would be invalidated. The same went for the density of the particles. Ehrenhaft assumed their density would be the same as the substance from which they were made, i.e. that the process of their formation didn't alter the character of the substance. Because these particles were considerably smaller than oil drops, Millikan also wondered if Brownian motions were interfering with the timings. If we approach the divergence of results between the two workers from the standpoint of the accepted physical theory of today, we can only conclude that Ehrenhaft must have been doing something wrong. Millikan rightly sought to give these suspicions a concrete form and to pinpoint the presumed error. Ehrenhaft, of course, and equally rightly, fought back. He actually managed to capture some of the individual microscopic particles he had been tracking and, using micro-photographic techniques, was able to check the circularity of their profiles. The result was that they looked like perfect spheres. He also performed experiments designed to ascertain directly the density of the particle material, to justify his previous assumption, and reported them vindicated. As for Brownian motion disturbance, he sought to overcome the problem statistically by taking large numbers of readings. Like Millikan, Ehrenhaft was worried about whether Stokes' Law would accurately describe the movement of his tiny particles through a gas. In response to this difficulty he utilized modified forms of the law that had been developed empirically to deal with small particles (see Ehrenhaft, 1941).

And so the debate went on, challenging and perplexing the scientific community for a number of years. Surveys and commentaries by competent and respected physicists reflected on the difficulty faced by the profession. Thus Lorentz said in 1916 that 'The question cannot be said to have been wholly elucidated' (Holton, 1978, p. 79); and in 1927 Chwolson said of the controversy: 'It has already lasted 17 years, and up to now it cannot be claimed that it has been fully decided in favour of one side or the other. . . . The state of affairs is rather strange' (Holton, 1978, p. 27). Holton is at pains to point out that 'there was never a direct laboratory disproof of Ehrenhaft's claims' (p. 79). No one, that is to say, got hold of Ehrenhaft's apparatus, or built a replica, and showed how by a greater exercise of experimental skill, or careful adjustment, you could get it to produce, and only produce, Millikan-type results. How, then, did the controversy end? How did the scientific community make up its mind – as it did – that it was Millikan and not Ehrenhaft who had succeeded in revealing a fundamental truth about reality? Holton tells us:

> Like most such controversies, this one also faded into obscurity, without anything as dramatic as a specific, generally agreed upon falsification taking place at all. Indeed Ehrenhaft continued to publish on sub-electrons until the 1940s, long after every one else had lost interest in the matter. (Holton, 1978, p. 79)

In other words, Ehrenhaft's findings gradually assumed the status of an isolated anomaly that simply didn't fit into the developing framework of current research and theory-building. It wasn't useful or used. From being a problem it became a perplexity, and finally a mere irritant.

Ehrenhaft's failure to concede the error of his ways came close to defining him out of the profession. The fact is that people simply stopped being prepared to listen to him, and conference organizers, for example, refused to give him a platform.

There is, of course, a simple explanation that suggests itself, one that would immediately clarify and justify all this. It is that Ehrenhaft was, and was known to be, a bad experimenter. This would explain why nobody bothered to furnish a decisive refutation of his claims, and why the controversy fizzled out in such a fashion: incompetence isn't worth spending time on. There are two pieces of evidence that might support this: one comes from the distinguished physicist Paul Dirac, the other from Holton himself (though Holton doesn't use it to press this point). Commenting on the dispute many years after the event, Dirac casts doubt on Ehrenhaft's competence: 'He was certainly not a good physicist' (Dirac, 1977, p. 292). Holton, for his part, points to the fact that Ehrenhaft's spread of results bears a certain resemblance to that produced by today's college students when they attempt to use a Millikan-type apparatus. They too tend to get a scatter of results which, if taken at face value, would also imply a unit of charge smaller than the electron. Let us take each point in turn.

A pronouncement by Dirac obviously has to be taken very seriously. It can be no part of the task of the sociologist of science to enter into a scientific dispute with such an authority. That is because the sociologist of knowledge isn't in the business of agreeing or disagreeing – that would be to do physics, not sociology. The task, rather, is to note and analyse and, if possible, explain such pronouncements. In this spirit we may notice two things. First, Dirac's judgement is retrospective. He had, in fact, been prompted into reflecting on the topic by a version of Holton's paper. Dirac didn't come to his conclusion as a result of watching Ehrenhaft in his laboratory, and having his suspicions aroused by what he saw. He reached his conclusion by an inference. Second, let us look at the reasons that underlie the inference. They are that Ehrenhaft's results show a wide scatter, and that most of the anomalous findings crop up with the smaller particles. This is said to suggest a lack of accuracy and a source of systematic error somewhere in the apparatus. In short, it adds up to bad physics. From our perspective the question that must be asked is how do physicists (whether they are a Dirac, a Millikan or an Ehrenhaft) decide between a genuine experimental effect and an error? If you are measuring a quantity known to have a stable and sharply defined value, and the measurements show a wide scatter, then clearly something is amiss with the technique of measurement. But if you don't know these facts about the quantity in advance, and if you are measuring in order to find out whether or not the quantity is stable and sharp, you can't a priori blame the measurements. The scatter might accurately reflect the range of values in reality; similarly, with the fact that the smaller sub-electronic charges tended to be associated with the smaller particles. This stands out as a systematic error if, and only if, there is a baseline of expectation that tells you that the addition and subtraction of charges works by way of units of electricity (namely electrons) that are independent of the size of the particle they are charging. But that was the very point Ehrenhaft was challenging. For him it was just an empirical fact, a law of nature as revealed by his experiment, that smaller particles attracted smaller charges. (He conjectured that it was connected with the smaller electrical capacity of the smaller particle.) It would appear, then, that Dirac is using another of Holton's 'plausibility arguments', i.e. using the outcome of an experiment, plus a taken for granted body of background knowledge, to deduce the general character of what 'must' have been happening. It was because Ehrenhaft didn't see that the 'natural interpretation' of his findings was systematic error that Dirac concluded that Ehrenhaft was not a good physicist (Holton, 1978, p. 292). We are therefore back with the familiar hermeneutics of the local interpretive tradition.

What about the similarity between Ehrenhaft's results and those of students doing Millikan's experiment in college laboratories? Isn't that embarrassingly strong evidence for the bad experimenter hypothesis? It might be, if it were not a priori unlikely, and if there were not other available explanations. Ehrenhaft, of course, wasn't a student, so it is somewhat unlikely that he would behave like one. He was a trained professional whose previous work was respected. We have already indicated that his first measurements had been taken up and cited by Rutherford. There were no qualms then about his competence. It was only when he began to publish deviant results that the question of competence cropped up. What appears to have happened is not that Ehrenhaft became careless, or lapsed into beginner's clumsiness, but that he changed his attitude towards the findings that were anomalous from the point of view of the electron theory. He began to see them as valuable and interesting, rather than as signs that something had gone wrong. The group of people towards whom he became oriented, and who provided him with an appreciative audience, were not, then, the Rutherfords and Millikans of the profession, but the anti-atomists and empiricists around Mach. He switched his allegiances and aligned himself with a somewhat different local, interpretive tradition, though one that was becoming increasingly beleaguered and isolated in Vienna.

The sociologist has little to say about such personal allegiances. No one knows where they come from, or why they may change. They are forms of personal choice, and the psychology of such processes is still deeply obscure. For example, we do not know why Millikan was so committed to atomism. In this regard he opposed the opinions of some of his respected and influential teachers. What makes one person accept, and another reject, the opinions of teachers and peers is unknown. But if we don't know the individual determinants, the sociologist can at least chart the options and describe the local, interpretive traditions that are available for the individual to accept or reject or utilize. These provide the resources and the standards by which individual contributions (and, ultimately, competence) will be assessed.

The suggestion is not that *all* imputations of incompetence have this form, viz that incompetence is detected simply by a failure to produce acceptable answers. All manner of tasks, such as timing the fall of droplets with a stop-watch, and remembering to check and correct fluctuating voltages, are of the kind that improve with practice. This is a known and familiar fact about a wide range of the skills that go to make up the conduct of experiments. Where the basic level of skill is known to be low, or where procedures and manipulations are unfamiliar to the experimenter, or where routine precautions are ignored, then we have independent grounds for suspecting incompetence. But such suspicions receive no direct support from anything that Holton or Dirac have said about Ehrenhaft.

In order to restore the balance of the discussion it is worth adding that Ehrenhaft was not always on the defensive. He too could point his finger at suspect procedures. For example, Millikan's apparatus could not handle as wide a range of particle sizes as could Ehrenhaft's own, nor could it be used under as great a range of pressures. Millikan was viewing the world through a smaller window. In this connection it is interesting to note that the very small particles observed by Ehrenhaft – what Holton calls 'ultra microscopic dust particles of metal' (Holton, 1978, p. 49) – were just the things Millikan sought to purge from his apparatus, because he found that they gave troublesome results (Holton, 1978, p. 319). Ehrenhaft's 'signal' was Millikan's 'noise'. Ehrenhaft also subjected some of Millikan's early published data to reanalysis on a particle by particle basis. He showed that when Millikan's interpretive assumptions were stripped away, the data assumed a scattered form more like Ehrenhaft's own. Millikan was here using the early water-drop version of his apparatus, but the crucial fact that Ehrenhaft uncovered was that Millikan had made arbitrary and inconsistent assumptions about how many electrons a drop was carrying. One droplet with a charge of 15.59×10^{-10} esu was credited with carrying fewer electrons than one with a charge of 15.33×10^{-10} esu. Holton described the attack as 'devastating' (Holton, 1978, p. 58). Nevertheless, no one seriously took it, or takes it, to impugn Millikan's general competence, and it was soon forgotten. Perhaps this is because the slip-up produced the right rather than the wrong answer.

How does all this bear on the criticisms that might be directed against the sociological reading of Millikan's experiment? We imagined a critic who asked: what can be 'social' about Millikan's results, given that anyone who sets up a comparable physical situation (who performs a relevantly similar experiment) gets the same results as Millikan? This tempts us to picture experiment as a kind of unproblematic looking. The experimenter will just 'see' certain things, and Millikan had found out, and correctly stated, what they will see. As in ordinary perception, the world just impinges on us, and society has no role to play in the process. The Ehrenhaft case reminds us that, as far as Millikan's work is concerned, it wasn't and isn't like that. Ehrenhaft, too, had set up his apparatus and established a routine way of interacting with the material environment. His apparatus also 'worked'. Like the potter and weaver, and like Millikan himself, he acted and intervened in the world, and received feedback of an intelligible kind from it. Like Millikan, and all other scientists, he exhibited a natural tendency to credit his results with significance, and to find plausible explanations which he was strongly inclined to believe were true. He was not moving in the realm of phantasy, or disconnecting his thoughts from reality any more than Millikan was. We could say that there was as much reality in the conduct of Ehrenhaft's experiment as there was in the conduct of Millikan's, if we concentrate on precisely what they did, rather than on how it seems in retrospect. There was just as much contingency underlying the way their apparatus behaved in the one case as in the other. Both scientists had their own explanations with which they sought to make sense of these contingencies and to explain, at least in outline, what their opponents must be doing wrong.

A scientific experiment as a whole, i.e. the detailed, real-life inter-action between the scientist and the material world, isn't a simple, transferable commodity that can be relied upon to reveal itself in the same form in Chicago, London, Vienna and Tokyo. Each experiment is subtly different. As Ehrenhaft pointed out, not even Millikan's close collaborators, who had become familiar with his experimental procedure over long periods of time, could simply go away and repeat his experiment in an unproblematic way. They often had to discard large numbers of results – sometimes up to 70 per cent of them – before they could extract the answer sought (Ehrenhaft, 1941, p. 443). The item that maintains its identity readily across time and space isn't the actual sequence of experimental actions, but the *proper result* of the experiment, and the associated standards for deciding if it has been properly performed. In short it is the experiment *qua* institution that is at issue here. And *these* items, the standards sustained by such institutions, are not instances of interactions between people, pieces of apparatus and the world; they are just between people, they are elements of human culture; they are *norms of interpretation*. It is the normative fact, the theoretically grounded notion of what a *properly* performed experiment of this kind must reveal, that lies at the heart of science's universality and repeatability (cf. Collins, 1992).

2.6 ANOTHER LOOK AT THE NOTEBOOKS

All the argument developed so far in this chapter has been prompted by reflecting on a single study. The conclusions we have drawn can be no stronger than their empirical basis. Like all historical studies Holton's work must be seen as a contribution to an ongoing, collective enterprise. Other scholars will go over similar ground, adding new considerations from adjacent areas or comparable cases, so the overall picture is bound to evolve and change. If other investigators were to discover specific grounds for challenging Holton's study the picture could change radically. There is, in fact, one such critical study currently available, and it is therefore appropriate to end this chapter with a look at this rival analysis of Millikan's work. Like Holton, Franklin (1986) has gone back to Millikan's laboratory notebooks but has reached significantly different conclusions.

The notebooks in question cover the work published in Millikan's 1913 paper in the *Physical Review*. In that paper Millikan calculated *e* on the basis of measurements taken from 58 charged oil drops. The notebooks, however, contain records for a total of some 175 drops. There is a slight discrepancy between Holton and Franklin here, perhaps because they have used different criteria for counting. Holton reports finding a total of 140 'separate runs' in the notebooks, while Franklin counts work on 175 'different drops'. The central question, as we have seen, is: what would have been the effect on Millikan's conclusions had he taken account of some or all of the excluded 117 drops? Was the discarded data independently identifiable as low quality or suspect, or was a significant proportion of it identified as unacceptable, not for independent reasons, but because it did not fit Millikan's presuppositions about electrons, their charge, and the non-existence of sub-electrons? Holton's conclusion was that Millikan's result, like Ehrenhaft's, was indeed sensitive to data selection (p. 69). Holton accepts that the inclusion of many of the discarded results would only alter the scatter, not the mean value of e. Nevertheless his case is that certain of the suppressed findings pointed to fractional charges, and hence 'sub-electrons'. Does Holton tell us exactly how many, and which, results were amenable to this reading? Was it 5, or 10, or 50 of the discarded results? Unfortunately he doesn't specify any such number. There is no statistical breakdown of the discarded data. Rather, we are given a striking, illustrative case of a sub-electron result which is analysed in convincing mathematical detail. The example is accompanied by the assurance that what applies in this case also applies to 'many others like it' (p. 68). The illustrative case was the second run made on Friday, 15 March 1912. This is the one mentioned above in our discussion. The measurements make sense if it is supposed that charges of $e/10$ are present.

Franklin, quite properly, tries to improve on this. He works systematically through the discarded readings in order to account for each in turn, and hence reach a precise estimate of their cumulative impact on Millikan's overall findings. Both his conclusion and his approach are strikingly different from Holton's. First, a remark about his approach. Franklin's overall concerns might be characterized as philosophical and normative rather than descriptive and explanatory. If Millikan has indeed been selective in the way Holton described then, as far as Franklin is concerned, Millikan has done something wrong: he has trimmed or cooked his data, and this is tantamount to fraud (p. 231). Like many philosophers and scientists, Franklin is prone to hear sociological description as if it were criticism. For Franklin the result of sociological work, including historical work by Holton, has been to 'cast doubt on the validity of experimental results' (p. 244). His general posture is thus one of defence, as if he were defending science against criticism. This defensive posture is, however, accompanied by a candid willingness to endorse such 'criticism' should any questionable scientific practices be uncovered.

Adopting this approach, what does Franklin find when he turns to Millikan's notebooks? His conclusion is that one experimental run – just one single result out of 175 – may have been improperly suppressed by Millikan in order to avoid giving Ehrenhaft ammunition. Interestingly, the one offending result is not the one cited by Holton, but rather the second drop studied on 16 April 1912 (p. 155). It was one of Millikan's best observations, showing all the internal consistency of a good run, and yet on calculation it indicated a fractional charge, or sub-electron. Millikan's jottings began enthusiastically with 'Publish. Fine for showing two methods of getting v'. Then the truth must have dawned, and on the bottom right-hand corner of the notebook were inscribed the words, 'Won't work'. Franklin suggests the drop may have been carrying a very large number of charges – (when more than 30 electrons were riding on an oil drop the apparatus didn't give reliable readings). Millikan himself didn't know this, and accepted no such limitation on his apparatus; so, in Franklin's eyes, he had no excuse for holding the data back. 'This', says Franklin, 'would seem to be a case of cooking' (p. 231). But it was only 1 out of 175, and Franklin assures us that it had no adverse effect on the overall course of science (p. 232).

Just as Millikan and Ehrenhaft made their experimental observations and reported conflicting things, so now Holton and Franklin look into Millikan's notebooks and their reports conflict. Holton concluded, though without any critical intent, that Millikan's results were sensitive to data selection; Franklin looks at the data and concludes that in general Millikan's results are not altered by selection – with the exception of one case, about which he is robustly judgemental. Has Holton simply got it wrong, generalizing incorrectly from an atypical case? What, indeed, has happened to the case Holton cited when it was subject to Franklin's re-examination? The conflict of interpretations is intriguing and needs probing into in more detail. Let us follow Franklin as he disposes of the 117 rejected readings of the notebook. Remember that 58 of the 175 entries in the notebook were published – this is why there are 117 to be dealt with.

First of all, Franklin disposes of all but 49 of them by arguing that they were never serious contenders for publication. They were, we are assured, readings taken in the course of testing the apparatus. Millikan obviously had to convince himself that his apparatus was working properly. All of the notebook entries prior to the date of the first drop that found its way into publication can be placed in this category. Now it might be that the notebooks clearly indicated such a cut-off point, but if so Franklin does not make this fact explicit. While some of it is certainly shown to have a 'just testing' character, the reader is left with the suspicion that this cut-off point is Franklin's own decision. He takes this to be Millikan's intent, and uses this gloss on the proceedings to justify the exclusion of 68 notebook entries.

This is a significant matter because discussions about when a piece of apparatus is 'working properly' can materially affect what is counted as the outcome of the experiment. Sociologists of science such as Collins (1992), Pinch (1986) and Pickering (1984) have paid considerable attention to processes by which a piece of apparatus is calibrated. Theoretical expectations and presuppositions are one of the factors that can insinuate themselves into this delicate and important process – and, of course, as far as they do then we have a version of the selectivity to which Holton was drawing attention. Unfortunately, in his discussion of Holton, Franklin passes over these considerations in silence.

Having effectively disposed of 68 drops on the grounds that they were test runs, this leaves Franklin with a remaining 49 excluded drops to be accounted for. They fall into two groups of 22 and 27 respectively which can be treated separately. The group of 22 are those for which Millikan took readings but did not bother to carry out calculations to arrive at a value of e; the group of 27 are those for which he did carry out a calculation, but did not use or publish the resulting value for e. Starting with the group of 22, Franklin reworks 10 of the cases to see their effect on the overall result. He concludes that their inclusion would not have altered Millikan's results in any significant way. As for the rest in this group he follows Millikan in discarding them for a variety of reasons. (One of those discarded by Franklin at this point is the run used by Holton to illustrate his case. We will need to return to this.) What of the 27 drops for which Millikan did carry out a computation? Franklin recomputes and accepts 15 of these in order to test their effect on Millikan's published conclusion. All the rest, bar one, he accepts that Millikan was right to discard. The exception is the drop of 16 April 1912 which Millikan appeared to have held back lest it give aid and comfort to the enemy – his one lapse from scientific probity in Franklin's eyes.

The upshot of Franklin's re-examination of the notebooks is to add 25 new results to the 58 published in Millikan's paper. The general effect of adding these 25 is primarily to increase the scatter in Millikan's findings, but not the mean value of the charge attributed to the electron. Taken at its face value this certainly has the effect of limiting the scope available for Holton's generalization from his paradigm case to the 'many others' said to be like it. After Franklin's work we can be sure that these 'many others' cannot include the 25 extra results Franklin has now brought into the fold. But though Franklin has certainly imposed some manner of constraint on the representative character of Holton's demonstration of sub-electrons hidden in the notebooks, it isn't easy to be sure of the exact character or scope of Franklin's criticism. The point at which one would expect the import of Franklin's criticism to stand out with maximum clarity would be in his handling of the very example on which Holton's case crucially depends – the discarded second reading of 15 March 1912, where Millikan appeared to have captured a sub-electron charge $e/10$. What did Franklin make of this?

Unfortunately, instead of being a point of maximum clarity in the conflict of interpretations, it is a point of maximum obscurity. One would have expected Franklin to have done one of two things: either (i) concede that there were really two small episodes where Millikan may have held back problematic data (rather than just the one episode Franklin has already conceded), or (ii) show that Holton's physics is wrong and that, after all, these readings are not amenable to interpretation in terms of sub-electrons of charge $e/10$. Perplexingly he does neither of these things. Instead he appears to take Holton to task for claiming something that he doesn't claim. Having identified the precise oil drop and the precise experimental run in question Franklin refers to the lack of internal consistency that provoked Millikan to write 'Error high, will not use. . . .' Franklin then goes on:

> Although there is a substantial difference between the results from the two methods of calculation, larger than any for which Millikan gave data in his published paper, it is no larger than the difference in some events that Millikan published, as shown in the laboratory notebooks. There is no reason to assume, as Holton seems to do, that there were unstated and perhaps unknown experimental reasons for exclusion of this event. (Franklin, p. 150)

This is the sum total of Franklin's discussion of the illustrative case central to Holton's argument. The oddity is that Holton didn't say, or imply, that Millikan excluded this event because of unstated or unknown experimental reasons (i.e. reasons like fluctuating voltages or convection currents). Holton said that if Ehrenhaft had been able to see this unpublished entry he would have been able to use it to support his sub-electron or anti-electron theory – and he then went on to demonstrate in quantitative, mathematical terms exactly how Ehrenhaft could have done this. Nothing Franklin says shows Holton is wrong. What is more, Franklin's concession that there were other readings with similar characteristics would seem to support Holton's claim that this reading is, indeed, not a unique or rare phenomenon.

Clearly the debate isn't over, and the situation is not a wholly satisfactory one. More work needs to be done. But enough has been said to reveal some of the difficulties and interest of using historical data of this kind. Perhaps the most striking feature of the episode is the way in which the interpretive process we first encountered within the workings of physics are replayed at the level of the history and sociology of scientific knowledge. If physicists like Millikan and Ehrenhaft can find themselves in the pull of competing interpretations of data, and find difficulty in resolving divergent perspectives, there should be little wonder that the same phenomena crop up in the historical sociology of knowledge. It is almost an a priori necessity that such difficulties as are encountered within the natural sciences will find their counterpart here too in the social sciences. If sociologists are to uncover *general* principles at work in the production of knowledge, these principles will inform their own attempts to detect, formulate, illustrate and justify those very principles. If they did not, then the sociology of knowledge would be in danger of demonstrating its own falsity.

CHAPTER 3

Words and the World

In the previous chapter we used a specific example to examine how scientists describe and interpret experience. We shall now look at the same problem in fully general terms. Our initial concern will be with how objects and states of affairs are classified, since to speak of things, in whatever way, is to speak of them as instances of some class or kind of thing.

Considered as a human activity, two images of classification readily come to mind. We can imagine an individual looking at natural objects and events, grouping like with like, separating what are not alike. Alternatively, we can think of the individual looking to what others do, or a teacher does, and grouping objects in the same way as them, classifying by acting as they do and identifying the classes produced with the same words that they use. In the first image, awareness of experience is the basis for classification; in the second it is awareness of the accepted practice of a tradition. Both images seem to capture something of what classification involves, but how are the two related? What are the roles of experience and tradition in the activity of classification?

One answer is that experience alone is the basis of scientific classification: the similarities and differences between things clearly evident to any individual are mapped by the terms in our language; tradition merely hands down this standard map and makes it immediately available to successive generations. As it happens, however, many alternative classifications of nature exist, sustained by different cultures and subcultures: there is no standard map. Nor do the members of these cultures and subcultures appear systematically to ignore the evidence of their senses, or the promptings of reason, in sustaining their own favoured version of the natural order. Hence sociologists have long been sceptical of the ability of experience alone to sustain any given system of classification, and have insisted on the essential relevance of the authority of tradition and of the way of life of the collective wherein the activity of classification is occurring. These seminal points for the sociology of knowledge were beautifully formulated by Emile Durkheim in *Primitive Classification* (1908, with Marcel Mauss) and *Elementary Forms of Religious Life* (1912). In effect, Durkheim treated classifications as social conventions, and alternative systems of natural kinds as alternative viable systems of conventions. But this did not mean that Durkheim considered nature to have nothing to do with how it was classified, and our experience of nature to be irrelevant to the growth of empirical knowledge. He merely held that it was insufficient:

> a certain intuition of the resemblances and differences presented by things has played an important part in the genesis of . . . classifications.
>
> But feeling the resemblances is one thing, and the idea of class is another. The class is the external framework of which objects perceived to be similar form, in part, the contents. Now the contents cannot furnish the frame into which they fit. They are made up of *vague and fluctuating* images . . .; the framework, on the contrary, is a *definite form*, with fixed outlines, but which may be applied to an *undetermined number of things*, perceived or not, actual or possible. In fact, every class has possibilities of extension which go far beyond the circle of objects which we know, either from direct experience or from resemblance. This is why every school of thinkers has refused . . . to identify the idea of class with that of a generic image . . . the best proof of the distance separating these two notions is that an animal is able to form generic images though ignorant of the art of thinking in terms of classes and species.
>
> The idea of class is an instrument of thought which has obviously been constructed by men. (1912, pp. 146–7)

Our objective in what follows will be to justify and to extend these claims of Durkheim. The strategy will not be, however, that of belittling the relevance of 'the world', and denying the importance of our 'intuition of the resemblances and differences presented by things'. On the contrary, the need is to give detailed attention to these very matters.[1]

Ostensive learning

All societies possess, as Durkheim implied, developed systems of classification, systems of terms which are used to speak of things as of this or that class or kind. Moreover, different societies sustain different systems, and classification for an individual in a given society is largely a matter of terms being drawn from the given inherited repertoire of that society, and put to use. It is true that human beings are capable of linguistic innovation, must be so capable if we are to understand the existence of human language in the first place, and it is true that such innovation may occur on any occasion at any time. But almost always people have recourse to inherited concepts and categories when they classify, and put these existing concepts to use to label whatever new objects or entities they encounter. In this sense all classifications of things are social.

Should we then think of our inherited system of classification as the 'framework', 'constructed by men', into which things and objects are fitted? This is sensible. How we classify will depend on which 'framework' we inherit and, because different systems of classification are inherited in different cultures, how the members of those cultures classify will be different: different conventions for classification will exist in different cultures.

But once we have the 'framework', will not attention to nature then suffice to fix how we use it? Is it not the existence of nature which accounts for people with the same framework using it in the same way? Is not this where 'the resemblances' come in: because we 'feel the resemblances' just as everybody else does, we fit things into our given framework just as they do? This is a beguiling conjecture, and it is most important to be aware of the grounds which make it necessary to reject it just as it stands.

First of all, it is easy to find examples which do not fit the conjecture. The taxonomies of the biological sciences, for example, are elegantly systematized hierarchical orderings of classes themselves, rather than of particular things (Hull, 1988). Hence, when these taxonomies are applied to particulars, nobody is surprised if scientists or groups of scientists find themselves having to agree to differ. At the opposite extreme, there are systems of classification which serve the practical purposes of scientists, and which apply to things, or kinds of things, only by virtue of what scientists conceive of doing with them: such are fertilizers, insulators, reagents and many other kinds of thing. Clearly, with classifications of this kind attention to nature will not suffice to fix how they are used, since what determines the use is not 'there' in nature in any case.

But although a sense of the inadequacy of the conjecture can be easily conveyed by examples of this kind, the way to expose its fundamental difficulties is to go for the hardest cases, cases where classifications seem to correspond to resemblances and differences in nature in a nice straightforward way, and their application to things and objects seems routine. This is why we shall begin with cases of classification that involve a minimum of explicit 'theorizing', and why we avoid instances of 'functional' classifications closely associated with highly specific modes of institutionalized activity. Whatever we find of sociological interest in our examples will also be present in the more common and complex cases alluded to above.

Let us take our everyday classification of the kinds of birds as such an example.[2] We routinely identify birds by their looks. How is this possible? It is, presumably, a matter of 'feeling the resemblances': ducks look like birds we have been taught are ducks; geese look like similar cases of geese; and so on. In acquiring the culturally specific framework[3] that is our classification of birds we have been shown examples of the different recognized classes and kinds of birds: this is what allows us to recognize further instances of each class or kind.

This characterization of how we acquire and become able to use our inherited system of classification will seem much too simple, but let us stick with it for the moment: complications will be attended to later. For the moment, we shall assume that classes are taught and learned entirely in this way, that is by *ostension*. Any act whereby a direct association is directly displayed or shown or pointed out between an empirical event or state of affairs and a word or term of a language will be called an act of ostension. Our assumption is that the inherited system of classification of kinds is learned by ostension.

This brings us immediately to a difficulty which will turn out to be of fundamental importance. No single act of ostension suffices to teach the correct use of a term: ostension is notoriously indefinite (Campbell, 1979). First of all it is never clear what precisely is being pointed to. There is the problem of separating field and background. One may point toward a swimming bird and say 'duck', and be understood as indicating the bird, or the lake it is swimming on, its beak, or even one's own outstretched finger. And secondly there is the fact that even when it is successfully identified, any thing is normally at once a member of several classes or kinds, and no single act will show which is being taught: to point out the bird as above may initiate indifferently the teaching of 'duck' or 'bird' or 'mallard'.

Fortunately, acts of ostension may be repeated for as long as is necessary to eliminate any *particular* source of confusion. Confusion with other kinds is lessened by pointing out more and more instances of the specific kind being taught, as well as by pointing to key examples which are *not* of the kind: 'this is a duck, yes, and this, and this, but no, [that is a moorhen], and no [that is a goose] . . .' Confusion with background is lessened by varying background, indicating a duck as it moves from lake to sky to land, indicating an absence of ducks from lake or sky or land when indeed there is such an absence. Confusion with fingers is lessened, if necessary, by use of thumbs or pointers. In practice, by pointing out a sufficient number of instances, a term like 'duck' can be successfully taught – 'successfully' in the sense that the person who acquires the instances becomes just as competent in the use of the term as anyone else. The instances of the term may be called a *similarity relation* in that they exemplify what it is to be 'the same' by virtue of being instances of the term. Thus, what it is about ducks which makes them 'the same' is exemplified by the instances in the similarity relation: this cluster of instances is all that ostensive learning can provide to convey that sense of sameness. Nonetheless, it may suffice to transform a learner into a fully competent member, able to recognize further ducks and label them correctly. Like everyone else the member will now be able to compare the appearance of future cases with the standard instances which have exemplified the similarity relation: those which resemble existing exemplary ducks, or resemble them more than they resemble non-ducks, will be called ducks: where resemblances go the other way, instances will not be called ducks.

We now have an example showing the most that can be achieved on the basis of purely ostensive learning, and we must spend some time reflecting upon its implications. Three points must be made, all of which are of crucial sociological significance. First of all, despite what is set out above, references to repeated ostension cannot provide a *formally* satisfactory account of classification. This is to be expected: if there is indefiniteness in every single act of ostension, then there will surely be indefiniteness in any combination of them. We can appreciate where this indefiniteness lies by considering what is involved in the notion of 'resemblance'. The next duck is classified as 'the same thing' as existing ducks because it 'resembles' them. Yet it is not manifestly identical to them: as empirical entities all ducks are different, just as all the members of any kind are different. The duck is the same as existing ducks, yet different from them: 'resemblance' must be a relationship of sameness yet difference. Unfortunately, resemblance of this kind will exist between any pair of things. Any thing may be said to be the same as, yet different from, any thing else. And, moreover, since

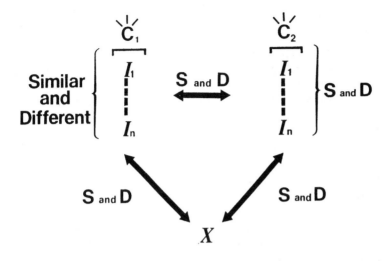

Figure 3.1

classification proceeds not on the basis of resemblance *per se*, but on the basis of greatest resemblance, so that things are grouped with what they *most* resemble, we have additionally to confront our lack of any *metric* for resemblance, any basis for a claim that sameness here outweighs difference there.

The formal situation is represented in Figure 3.1. The teacher uses ostension to present two classes of things, C_1 and C_2. They may be 'duck' and 'duck', or 'duck' and 'goose', or any other set of terms at all. The various Is are specific ostended instances. X is the next instance, the instance just encountered, not yet classified, about to be classified. The figure simply dramatizes the fact that resemblances link all instances. Considered formally, ostensive learning offers us *nothing* by way of guidance in the classification of the next instance. Any classification of the next case can be reconciled with what has already been learned ostensively; however other things have been classified so far, the next thing can be classified in any way without any formal inconsistency with earlier practice. Thus, the form of the figure will serve as a useful mnemonic, a symbol of a fundamental problem which will recur in a number of guises.

The second crucial point is that this general problem of ostensive learning is not merely a problem of purely ostensive learning. All attempts to classify things on the basis of their empirical characteristics encounter the same problem, even if the relevant terms have not been learned wholly by ostension. Suppose that we allow that people may be *told* how to use a term as well as *shown*, so that the term can be given a position in a *network of beliefs*. Clearly, anyone who understands the place of a term in such a network will manifest that understanding in his or her use of the term. To be aware that ducks have webbed feet, or that teals are ducks, will affect the way that 'duck' is applied. It may even be that ducks are held to be *by definition* kinds of web-footed birds, or that there is a *rule* to follow, requiring ducks to be web-footed. How far might rules or definitions serve here to eliminate indefiniteness, and fix how 'duck' is to be applied?

The practical value of definitions and rules in specifying more precisely how terms should be applied is not in question. But they cannot solve the formal problem. A definition or rule could only solve this if its constituent words were themselves devoid of indefiniteness in use. But how are such words themselves learned? If by their role in other word–word connections, then we enter upon a vicious regress; if by ostension, then we lapse back into indefiniteness once more. Having 'defined' duck by reference to web-footedness or whatever, we still are less than fully clear how to use the term because we are less than fully clear about what counts as web-footed. A rule for identifying ducks would give rise to the same difficulty.[4] In the last analysis, the source of empirical significance is always a word–world relationship; word–word relationships merely shift the place at which a link between words and the world needs to be made. The 'complications' referred to earlier stemming from the fact that classifications are not learned *wholly* by ostension turn out not to be material to our key conclusion. At some point we have to stop talking about word–world relations, and simply *show* them. As T.S. Kuhn has put it, there is always a point where we have to recognize that some things are the same as others, without being able to say in what respect they are the same.

Ostensive learning of classifications implies indefiniteness in their future use. Yet all empirically applicable terms must be learned, directly or indirectly, by use of ostension. These are our first two fundamental points, and they are arrived at by formal argument. The third and final fundamental point is, by way of contrast, an empirical generalization. Despite the formal limitations of ostension just outlined, it is nonetheless the case that classification exists as a routine accomplishment in all communities, and that ostensive learning is a key process through which that accomplishment is actually sustained and transmitted. The fascination of ostensive learning derives from this contrast between its problematic and opaque character when considered formally, and its ubiquity and general straightforwardness when considered empirically. Everywhere, in all societies, ostensive learning is successfully accomplished. Typically, it is easily accomplished and transmits not mere competence in classification but an amazing *facility*, so that words are attached to things not just appropriately, but unproblematically, immediately, without thought, as a matter of course, in the vast majority of cases.

This amounts to an argument for attributing certain minimal *inherent* competences to individual human beings. They must have an inherent sense of *sameness*, so that initially each spoken sound 'duck' is recognized as 'the same' as the next, and later each yet to be encountered duck is seen as like earlier exemplary ducks. They must have some inherent awareness of association, such as allows the proximity of a word and a thing to be treated as a connection. And they must have an innate *inductive* propensity, a disposition to accept past experience as relevant to present circumstances and past associations of words and things as persisting associations.

Sociologists tend to avoid conjectures about individuals, and especially nativist conjectures. Their accounts of classification tend to appeal to culture rather than nature, to learned competences and proclivities rather than inherent ones. But these things need to be run together: it is to the symbiotic operation of nature and culture that we must refer, if classification as human activity is to be adequately understood. Indeed at this point we must return to Durkheim, and reject his belief that the world itself is incapable of providing any stable and distinct patterns to an unsocialized mind: Durkheim's contrast between the 'vague and fluctuating' images of perception and the 'definite form' of the cultural framework is a false one. Our perceptual apparatus is our only means of access not just to nature, but to culture as well. That the required sense of what is the same and different in nature may be successfully conveyed by the use of the words of our culture is evidence of innate, pre-social, perceptual powers. That we are aware of a cultural framework with a 'definite form' implies our capacity to apprehend 'definite form' in nature.

A finitist account of classification

To discuss ostensive learning is to discuss classification in a way which gives the maximum possible priority to the role of empirical experience. It enables us to return to our key questions concerning the relative importance of experience and tradition, the relationship between 'feeling the resemblances' and 'the idea of class'. Notice that ostensive learning is itself a social process. The learner is taught how to grasp things by other people, not by the things themselves, which remain silent and unconcerned. The right way to grasp things is established as convention in the tradition, and is transmitted in a social relationship involving trust in the teacher and acknowledgement of his or her cognitive authority. Which right way is taught will depend upon the tradition in which the learner is embedded: we considered learning about birds; in another context there might be no class of feathered animals to learn, but instead a class of flying animals, including, as we would say, birds and bats;[5] in scientific subcultures learning might be based upon sameness relations different from either of these. Each tradition orients us to experience in its own way: in its similarity relations, its own particular cluster of exemplary instances of its terms, it shows us what it counts as being *the same thing*. It gives us a specific conventional terminology with a 'definite form', a 'framework' into which the 'contents' of experience will fit.

Once in possession of the conventions of their tradition members will apply them and perpetuate them. Knowing what counts as 'the same thing' they will go on and themselves identify further cases of 'the same thing'. But there is an ineliminable indefiniteness associated with this 'going on'. Members cannot simply be given the conventions and left to 'go on' as though there is a railway track stretching out ahead of them. It is not the case that, once they have the conventions, 'feeling the resemblances' will suffice. A resemblance is not the same as an identity. It offers grounds for claims of sameness and claims of difference, and no indication of which should have priority. To apply a term on the basis of resemblance is to extend an *analogy*, a process the proper performance of which cannot be adequately specified. In the last analysis, whether such an analogy is made will depend upon whether or not people are disposed to make it. And how classifications develop in the future will depend upon how people will be disposed to develop them in the future. In a nutshell, future use of our conventions of classification is underdetermined and indeterminate. It will emerge as we *decide* how to develop the analogy between the finite number of our existing examples of things and the indefinite number of things we shall encounter in the future. This view of classification is sometimes labelled *finitism*. Its importance for later arguments is such that it is worthwhile to list its central claims, and some of its more significant implications systematically.[6]

1   The future applications of terms are open-ended.

This is merely a reformulation of the central tenet of finitism. On a finitist account there is nothing identifiable as 'the meaning' of a kind term, no specification or template or algorithm fully formed in the present, capable of fixing the future correct use of the term, of distinguishing in advance all the things to which it will eventually be correctly applicable. In a sense, we never know what our kind terms mean, however readily and unproblematically we apply them, for we never know how they are going to be used. The use of our terms is *open-ended*; there is no definite class of things to which they already apply, no closed domain of application the boundary of which can presently be discerned.

From a sociological perspective the importance of this key theme of finitism is that it reminds us of the status of every act of classification as a separate and problematic empirical phenomenon. It sustains and intensifies empirical curiosity, particularly with regard to the agreement that members normally achieve in their classificatory practice. It highlights the conventional character of classification and the need to understand convention as a collective accomplishment.

This idea can give rise to difficulties. The notion that classifications are conventions is widely accepted, as a result of our experience with alternative ways of classifying the world, all manifestly sensible and satisfactory. But we tend to think of the associated problems simply as those of choosing conventions. We imagine that once they are chosen they will then determine our subsequent taxonomic activity. It is as if the world is a cake, ready to be cut in any number of ways, indifferent to how it is actually sliced. We decide on how to do the slicing, but once it has been done the status of everything in the world is fixed, and we must subsequently proceed as the conventions 'require'. Reflection on the empirical process of classification suggests a better metaphor. We have put our knife into the cake and cut a certain way. But nothing determines how we should continue to cut: we do not have to cut in a straight line. And indeed there is nothing to stop us pulling back the knife some way and starting again. In so far as classification is conventional, it is our selected ways of applying our terms which will determine what form the conventions have and will come to have, not the conventions which will determine our ways of applying terms. Some ways of cutting may come to suit us better than others, given the way the world is, but the cutting continues indefinitely, and the knife remains always in our hands.

2 No act of classification is ever indefeasibly correct.

This is another well-rehearsed point. A nice way of presenting it is as an implication of the fact that classification can never proceed on the basis of *identity* of appearance, and has to proceed on the basis of *analogy* in appearance. There is no correct or incorrect way of extending an analogy which can be read off, as it were, from the appearances themselves. People must *decide* what is correct and what is not. From a sociological perspective this usefully highlights the role of collective judgements, negotiated on the basis of a range of more or less compatible individual perceptual intuitions, in defining what will count as correct classification.

None of this is to assert that an individual cannot have a compelling conviction that just one way of applying a classification is correct, nor that such a conviction may not be shared by an entire group of individuals as a result of separate experience. It is indeed the case that most of the time routine, matter-of-course, unproblematic use of classifications is the order of the day, and that an unreflective agreement in practice gives a clear indication of what is correct without any need to 'achieve' consensus. The 'only right way' to extend an analogy may be identified with consummate ease – so readily indeed that the process appears automatic and machine-like; this is simply a matter of fact.

Another matter of fact is that analogies may at times be extended only with difficulty, and that competent individuals may differ in what they find the most sensible and appropriate way of proceeding. When this is the case the judgemental character of classification may more readily be recognized. The experience of difficulty may make it easier to appreciate Wittgenstein's tendency to treat classification as creative activity, or Garfinkel's proposal that whenever a term is applied, it is applied for 'another first time'. What needs to be emphasized however is that the blind, unhesitating, routine use of a term and its creative, imaginative, consensually sanctioned extension are extreme manifestations of a single kind of operation, in which a classification is applied to the next case by analogy with existing ones.

3 All acts of classification are revisable.

The finitist view of an act of classification as a contingent judgement implies not just the open-ended and indeterminate character of future acts, but the provisional, revisable character of previous ones. Any act of classification, and hence any pattern or system of classification, may be held to have been mistaken. Another way of extending the analogies involved may always be held to be the 'correct' way. In classifying we reserve the right, as it were, to look back and declare that we have done it wrong. Even among our standard instances of ducks may lurk, so we may come to decide, significant 'mistakes', instances which are 'not really ducks' after all.

From a sociological perspective, this serves as a reminder that the authority of collective judgement extends across the whole of time, to the past as well as the present. The terms 'dead' and 'alive' may be successfully applied on the basis of simple empirical examination, and indeed dead persons are routinely discriminated from living persons on this basis. But medical practitioners who make this distinction as a matter of course in their day-to-day activity may at the same time reflect back upon their own practice, whether in particular cases or in general, and find it wanting.[7] They may conclude, for example, even if at first only in their closed conferences and weekend meetings, that some bodies with the appearance of being alive and routinely treated as 'living' in existing practice, are 'really' dead and better treated as such. In concluding thus, practitioners typically hold that bodies of the kind in question are not merely being incorrectly identified now, but have been in the past. The conclusion is taken to have implications for earlier practice and for the way that instances were assimilated to similarity relations in that earlier practice.

Classifications in the natural sciences are constantly being revised. It is the business of scientific research to modify existing knowledge, even as it relies upon it. Even more than in everyday life, the revision and reformulation of what is known is a ubiquitous feature of normal practice. We tend to underestimate its extent and importance.

4 Successive applications of a kind term are not independent.

To use a kind term, to apply it to a thing, is to add the thing to the existing similarity relation and hence to change that relation. Every bird which we classify as a duck becomes an additional unique member of the set of examples which is the basis of our understanding of what a duck is. We could say that to use a term, even in the most routine way, is to change the meaning of the term; that it is impossible to represent classification as so many independent acts of use of a term, all isolated from each other, all deriving separately from the term's meaning. But perhaps it is better to eschew reference to 'meaning' altogether, and to say simply that any act of use of a term is liable to condition all subsequent acts of use.

A standard example here invites us to start with two tins of paint, say a blue and a yellow tin. We then mix the paints appropriately to produce a range of intermediate shades from one pure form to the other. Finally we increase the number of intermediate gradations until any adjacent pair of colours is indistinguishable one from the other: where difference is discernible we simply create more intermediate shadings as necessary. Now it is obvious how routine, matter-of-course extension of a same-colour relation can result in its radical transformation.[8] The example is striking in that the analogy of appearance involved here may actually be a matter of *perceived identity* of appearance, the strongest imaginable empirically-based impetus to applying a single term to a series of things. Yet direct comparison of the beginning and end of the series will immediately undermine any sense of indistinguishable characteristics being grouped together. (The example also serves beautifully to reveal the dangers of legitimating assertions of sameness by reference solely to the strength of immediate intuitions thereof.)

As a term is used, so its subsequent use is conditioned. One act affects another. In a collective, terms are applied by different individuals at different times in different contexts: the exemplary instances of proper applications of a term are collectively established. In a collective, therefore, where terms are shared as part of the language of the collective, what one individual does will affect what another subsequently does with a term. From a sociological perspective, the interdependence of different acts of classification is a way of describing a form of interdependence between people.

5 The applications of different kind terms are not independent of each other.

If we seek to classify on the basis of resemblance, we must operate in terms of maximum resemblance: the next instance belongs amongst those existing instances which it is 'most like'. This means that the next instance must be compared with instances of *all* the relevant kind terms before it is classified into that kind where the greatest analogy is perceived. To be fully competent in the use of any term, therefore, a member must be competent in the use of the whole system of classification: he or she must be aware of relevant instances of all existing kinds. No system of classification is so many separate, independent pieces; any such system must be transmitted and acquired as a coherent whole. And the authority of the teacher(s), correspondingly, must be an authority extending across an entire system of terms, a whole area of culture.

Consider too that every act of use of a kind term changes the associated similarity relation (4 above), which relation is implicated in the application not just of the term itself but of all associated kind terms. It follows that any act of use of a term is liable to condition all subsequent acts of use of all those associated terms. It is not just that how we apply 'duck' today is liable to affect how we apply 'duck' tomorrow; it is also liable to affect how we apply 'goose'. The activity of classification must be understood holistically. Similarly, in a collective, how some individuals use 'duck' may affect how others use 'goose'. A holistic understanding of classification must be a sociological understanding: developing interactions between the uses of terms will be identifiable empirically as developing interactions between people.

3.2 VARIETIES OF EXPERIENCE

Finitism is an attempt to describe classification as human activity, and to raise problems about it as an empirical phenomenon. But, so far, although it has been presented as an account of any system of classification based upon empirical information, it has only been exemplified by, and hence shown to be plausible for, terms applied on the basis of immediate appearances. There is more to classification than a concern with empirical information, as the next chapter will describe, and there is more to empirical information than appearances. Certainly, the empirical sciences are concerned with far more than appearances, even if we include amongst appearances the image provided by a telescope, the sight of a metre-needle against a background scale, or the lines drawn by the pen of a chart-recorder. Natural science, like everyday discourse, speaks of an enormous variety of empirical experience, and this must be properly allowed for. The argument in what follows is that it can be allowed for without the need for any significant change in the finitist account.

Complex elements of experience

We have discussed ostensive procedures mainly in terms of bounded, topologically simple objects and their surface features. But similar procedures can be used to build up word–world relations with any elements of experience that human beings have the power to discriminate and apprehend. Ostended objects do not have to be material objects; they do not have to be topologically simple. The features or properties of objects do not have to lie on their surface or even within them. Consider holes or boundaries: these things are readily shown and classified. Indeed similar objects are routinely classified in the sciences: cavities and their kinds are of deep interest in the physical sciences; unconformities are objects which are only too easy to bring to the attention of the geology student on a field trip; consider too the valleys, summits, ridges, cols and corries which may interest the cartographer or geographer as much as anyone else. We have a remarkable ability to comprehend spatially extended pattern, allowing a great variety of forms of ostensive learning, sustaining a correspondingly vast number of possible classifications. But it is clear that, with all the kinds of things thus classified, the finitist account will continue to apply. The problem of the next hole or the next boundary is analogous to the problem of the next duck.

Just as we can directly apprehend pattern in experience extended across space, so we can apprehend it in experience extended across time. *Processes* can be ostended and made the basis of classification. Thus, things can be classified as hard or soft materials, refractories and non-refractories, bleaches, explosives, and so on. Moreover, these classifications could be taught directly by ostension and naught else, even if in practice they rarely are. The difference between hard things and soft things could, for example, be established as a primary distinction of kinds of thing, purely by reference to processes like scratching or bending. There is not, however, a perfect analogy between our experience of spatial and temporal extension. Unlike space, time is recognized as having a direction, so that whereas only associations exist in space, sequenced relations of things are recognized in time: some things may be recognized as the causes of others, and classified thereby. Thus we have antigens, enzymes, hormones, toxins and so on, all classes wherein things are ordered by their 'causal powers'.

Once it is recognized that processes can be directly ostended it becomes much easier to understand how the finitist account can be extended to the physical sciences. Many of the key concepts of the physical sciences are taught in an abstract manner which makes it hard to perceive any analogy between them and the terms for kinds of things: 'space', 'time', 'distance', 'speed', 'mass', and so forth seem altogether divorced from notions like 'duck' or 'goose'. Yet the physical terms are the terms of an empirical science. What then is the nature of their link with experience? What word–world relations are involved? A plausible hypothesis is that these terms represent refinements of everyday notions taught and exemplified by direct reference to simple, physical processes.

Consider the term 'speed'.[9] References to its definition in current textbooks merely connect the term with other terms – 'distance', 'time', etc. – which themselves need empirical elucidation. Speed, and the terms with which it is associated, must in the last analysis be understood through examples. The movement of objects against a background is presumably one form of experience which would permit direct ostensive learning of 'speed', and the 'same-speed' relation. But many examples of the process would have to be involved: 'same-speed' conceptions extend not just to regular straight-line motions but to curved trajectories, erratic motions, motions in systems which are themselves in motion, motions of composite objects like armies and fleets. In acquiring competence in our common culture we learn 'speed' much as we learn 'duck', by familiarity with many actual instances of the use of the term. The finitist account must be expected to apply, and so it does. No great amount of reflection is required to conclude that we no more know what 'speed' means than we know what 'duck' means.

Why, though, do things appear to be different in the physical sciences? Why does the analogous term 'velocity' seem so much more refined and precise than 'speed'? When we get to the heart of the matter here we shall find that the differences between the scientific and the everyday notions are interesting but not at all fundamental. First of all we must set aside problems of quantitative measurement as incidental, and recognize that the real problem here is to identify what it is that is being measured: the driver with the most advanced speedometer is not necessarily the one with the best understanding of the notion of speed. Secondly, we must recognise that 'velocity' notions in science tend to be made derivative of notions of 'distance' and 'time', so that some of the problems of its empirical exemplification pass over to these notions and tend to get lost in the distance. Finally, in the sciences examples of terms like 'velocity' become broken into more elaborate groupings, so that many of the obvious differences between instances in the initial 'non-scientific' notion of 'speed' no longer remain unverbalized: the physical scientist recognizes distinctions between absolute and relative velocity, linear and angular velocity, average and instantaneous velocity, and is able to highlight the differences involved by direct reference to specific examples and illustrations which would all alike be liable to fall under an undifferentiated lay notion of 'speed'.

Nonetheless, the finitist account continues to apply. Clusters of instances underpin our understanding even of these elaborately differentiated terms. They appear precise in comparison with analogous everyday notions, and indeed they are more precise in the sense that they allow more differences and distinctions between cases to be verbalized. But they must always remain cluster-concepts nonetheless, concepts exemplified by non-identical paradigmatic cases of use. Or, at least, they must do so unless it happens that after a certain amount of teaching the human mind abstracts 'pure ideas' from particular cases, ideas of 'velocity', 'time', 'length' and so forth, which, once grasped, account for all further applications of the associated concepts.

The history of science offers support for a finitist conception of the terms of physics and mechanics. In a brilliant discussion of 'A Function for Thought Experiments' (1964), Thomas Kuhn notes how major advances in scientific fields may amount to a recognition of important differences between examples within what is initially a single similarity relation. Thus, Aristotelian physics failed to distinguish between what we might now call 'average speed' and 'instantaneous speed': these notions were themselves unverbalized and not thought of. Instances of both were clustered together under a single similarity relation for 'speed'. Galileo was responsible for the instances originally grouped in one relation of sameness being separated into two. But he did not do this by calling attention to the lack of *identity* between different examples of 'speed', which would have been mere pedantry. Instead, he used a thought experiment to show that the two hitherto unremarked kinds of example of 'speed' could yield incompatible conclusions in their actual application, that it was important to separate them for practical purposes. Galileo thought up a physical situation involving objects A and B such that there was good Aristotelian precedent for saying of the situation that A was faster than B (instantaneously), and also good precedent for saying that B was faster than A (on the average). Since the situation he thought up was of a kind that could well be encountered in practice, Galileo thereby demonstrated the expediency of splitting Aristotelian precedents into two distinct sets. Science was improved for practical purposes without additional empirical investigation by reflection on the clustering of instances under terms, and the possible reordering thereof.

It is true, of course, that however many times the finitist account is persuasively exemplified, its further extension can always be resisted. Thus, it may be claimed that the history of science merely records the process whereby scientists developed their notions into their final 'pure' form, where they at last came to correspond to 'pure ideas'. On this view, finitism describes science wherever it is 'not quite finished'. How can one oppose such a view? One can only point to further examples, and use them to cast doubt on unduly optimistic intuitions about the clarity of our terms and concepts: where we have a single word, we too easily imagine that we have a 'pure idea'; when we use the word we too easily imagine that we know perfectly well what we mean. Einstein may, in retrospect, be said to have clarified our notion 'mass', by distinguishing between the notion of 'inertial mass' and 'gravitational mass', but many found the notion of mass as clear as crystal beforehand; indeed the physical theory in which the undifferentiated term had a role was long regarded as a model of clarity and precision, despite its conflation of these two quite different notions.

Mathematics offers some of the most striking examples here. For all the apparent simplicity and transparency of the more elementary terms in mathematics their peculiar difficulty is quickly apparent to anyone who seeks to teach them. Problems are soon encountered because of the way that the cardinal and ordinal aspects of the use of numbers run together unremarked; because a minus sign serves sometimes as a property of a number and sometimes as an operator; because of the perversity of zero as divisor or exponent, and so on. Under the unity imposed by mathematical terms is an immense diversity of exemplary applications essential to attaining competence in the use of the terms, exemplary applications which are in no way intelligible as manifestations of a single 'pure idea'. Yet it may well be that if the teacher admits of no difficulties, then neither will those she teaches. It is a matter of history that generations of children have routinely learned the standard formulations of mathematics from texts full of formal confusions and logical inconsistencies.[10]

A circle is commonly defined as the locus of a point which moves in a plane yet remains at a fixed distance, r, from a second point. Having learned this, students may go on to learn, perhaps in the next lesson, that the area of a circle is πr^2 and its circumference $2\pi r$. They will go on to understand how a circle can cut another circle at not more than two points, or an ellipse at not more than four, and how it can itself be cut into sections and segments by its radii and by chords. The point of mentioning this chaotic and contradiction-ridden area of mathematics teaching is not to denounce its inadequacies, but rather to call attention to its adequacy, and to the fact that it is scarcely ever perceived as defective. Class after class of students have sat, as we in our time sat, learning of circles for weeks or months, sometimes as discs, sometimes as rings, with incompatible piles of information accumulating. And never an eyelid has batted. People have apparently found little problem in accepting that a circle is a line, and that notwithstanding it has a finite area. Nor have they routinely come to realize that in learning of circles they could not have been learning of 'one thing', let alone a 'pure idea'. Teaching has ordered the various examples of 'circle' so that no difficulties have arisen in the use of the term in practice; and in consequence the single term 'circle' has concealed the lack of any underlying 'pure idea' with quite staggering effectiveness. What could serve as better illustration of the need for caution on the part of those who would limit the scope of the finitist account?

Spatio-temporal continuity

Our empirical information about a thing concerns not just appearances and properties, but its place in a spatially and temporally extended world and its relationship with other things in that world. This kind of empirical information can be used as a basis for classification as well as, or even instead of, the sort of information we have considered so far. In particular, sameness may be established on the basis of 'same-place' relations. Whatever occupies a given region of space over adjacent instants of time may be counted 'the same object': sameness may be inferred on the basis of this form of spatio-temporal continuity. This topic has been treated with wonderful penetration by Saul Kripke (1972).

Kripke begins by asking how proper nouns are to be correctly applied, and what they are to be taken to refer to. For his paradigm example he takes the name of a particular human being, say John Brown. What is being referred to by the words 'John Brown'? Kripke's first point here is that appearances have no crucial relevance to the correct application of the name: different manifestations of John Brown are not 'the same person' because of appearances, or properties, or sets or clusters of properties. The young John Brown and the old John Brown do not count as such on the basis of any relation of resemblance at all: resemblance is in the last analysis incidental to correct use of the name. However John grows and develops, whatever accidents befall him, whatever traumas afflict his body, he remains John Brown. Nor can any empirical property or propensity be specified as necessary for John to be John Brown. Even behavioural properties are incidental to the identification: an insane John Brown is just that, still John Brown albeit insane. Similarly, we can contemplate a comatose John, an amnesiac John, a hypnotized John. No empirical variation upon John serves to destroy his identity as John, or so it appears. And yet, quite rightly, we sense that there is an empirical basis for our use of names. What then can it be?

Kripke's answer is that our sense of spatio-temporal continuity is what sustains our naming activities. A bounded object arrives in the world in a certain context, in a certain way, which entitles it to be named as a particular human being: thus the object that will be John Brown arrives and is duly 'baptized' as John Brown. From that point on, whatever object is in continuity with the initial John Brown is *ipso facto* itself John Brown: that object which is roughly congruent with the original John an instant later is itself John, and so on from instant to instant – unto the last syllable of his recorded time, as it were. No matter that the initially baptized object is continually changing. The changes are changes in John, not from John into something else. All the empirical characteristics of John may change to any extent, for John can still be designated John on the basis of his existence as a topologically simple object in space–time. Kripke calls this *rigid designation*. He argues that people generally recognize terms like John Brown as rigid designators, and that this is evident in their mode of use: 'John Brown' is taken to refer to a soul or essence, to some permanent unchanging element which remains unaffected by all the manifest changes and developments in John Brown's body. By this 'metaphysical' mode of speech people downgrade the importance of one set of empirical information (concerning appearances and properties) in relation to another (concerning spatio-temporal continuity).

But if appearances are irrelevant to John Brown's being John Brown, if spatio-temporal continuity is the sole basis for the conviction that he is still there, then must we not monitor John 24 hours a day, with unswerving attention, to continue to know who he is? Kripke's response is to concede that appearances may indeed be relevant to keeping tabs on John, but that appearances only have the role of *indicators*. Although John may at times be *picked out* by his appearance, he *is* that which is continuous with his earlier manifestations. Moreover, whilst an individual may never be in a position to monitor John on the basis of continuity, a *collective* may come much closer to successful achievement of the task. Naming must be thought of as fundamentally a collective accomplishment. A collective may make continuing references to John Brown on the basis of spatio-temporal continuity. An individual member may step in and out of the collective linguistic activity, using the speech of fellow members to relocate John Brown, and may thus continue to relate to him as a continuing object in space–time. John becomes tagged, as it were, with his name: the name, in the speech of those surrounding him, indicates that he is John, rather as a label tied to a rose may indicate that it is a shrub or a climber. Individuals returning to the vicinity of John may use the speech of fellow members as a tag to reidentify John and speak of or to him: once back their speech will serve as a part of the tag used by other individuals. John becomes identified and reidentified by empirical features of the context *outside* of him, the social context in which he exists and has standing.

Kripke's discussion of names and their references offers an account of how we identify a specific human being, but Kripke himself recognized its possible relevance in understanding classifications in the natural sciences. He suggested that the terms used by natural scientists to refer to *natural kinds* have a close kinship with proper nouns. And the suggestion has been taken up and built upon by the philosopher of science Hilary Putnam (1975).[11]

Natural kinds have a special standing which marks them off from the main run of kinds or classes. Their instances are held to be separated from other things by a clear natural discontinuity: there is no question of their being demarcated from a continuum merely for practical convenience, like size-three eggs or second-early potatoes. And it is to instances of these kinds that the laws of the natural sciences are held to apply, and to apply straightforwardly and unproblematically because of the sharp 'natural' boundary defining the kind. Most of the basic scientific disciplines are engaged with the investigation of their own characteristic natural kinds: examples include the fundamental particles of physics, the elements and compounds of chemistry, the species of the biological sciences, the bodily organs and structures in anatomy and physiology.

Kripke and Putnam are right: the analogy between natural kind terms and proper nouns is very strong; Kripke's valid insights into our use of the latter begs to be extended to the former. Thus, natural kinds are not entities which belong together because of likeness of appearance or properties. Diamond and carbon black are both unquestionably elemental carbon, yet the two could scarcely differ more to the eye, or indeed in their properties and behaviour: even their 'chemical' behaviour is systematically distinct, and will not unproblematically legitimate their being taken as examples of 'the same element'. We may on occasions *recognize* carbon by appearance, say in the form of diamonds, but it is not that which allows us to recognize carbon which makes the carbon carbon, or else carbon could only be diamond and nothing else, which is false. Notice also that nothing in science prohibits the existence of some as yet unknown allotrope of carbon, some further form with properties as yet unascertained. The mere recognition of this possibility surely establishes that it is not a relationship of similarity or resemblance which permits us to decide what the next case of carbon will be. Similar considerations could be adduced with any element, and the argument is readily extended to natural kinds generally.

On Kripke's account, instances of natural kinds will be identified on the basis of spatio-temporal continuity. Black carbon can be squeezed into diamond with sufficient pressure: where the one was before the squeezing, the other is found afterwards, another manifestation of 'the same thing'. The case is even easier to appreciate with regard to the natural kinds of biology. Species terms are often considered as paradigms of classification on the basis of observable empirical properties and appearances, but this is not at all the case. The caterpillar goes into the chrysalis and emerges as a butterfly: its appearance and capabilities are mightily transformed, but its species designation remains unchanged. Spatio-temporal continuity between the instances makes them 'the same'. The instances of species are not in the last analysis specific things as they exist at specific times: they are life cycles. Biological communities keep tabs on instances going through life cycles and set whatever appears at any point in a same-species relation with whatever appears at any other point.

But if Kripke and Putnam offer valid insights into natural kinds, how far are those insights consistent with finitism? One way of reconciling Kripke with finitism would be to suggest that he has merely identified an interesting *option* which people often take when they apply terms and develop usage (Barnes, 1982b). They may indeed usually speak of the insane John Brown as precisely that, but where is the rule which prevents them agreeing with John's distressed mother that 'he isn't John any more; he's someone else'? Why should they not decide to take account of appearances and empirical properties in preference to spatio-temporal continuity, on some occasions if not on others? The question is a good one, for language users do indeed sometimes develop usage on the basis of appearance and sometimes on the basis of continuity, even when the terms are natural kinds. There is indeed a botanical speciality, numerical taxonomy, which seeks to identify and differentiate species wholly on the basis of appearances, and which holds an appropriately consistent view of 'what species really are'.

If Kripke has merely identified a neglected option available to us as we extend a term from one thing to another, then he has not undermined a finitist account but enriched it. We are now aware of yet another basis upon which one thing may be held 'the same' as another as a matter of contingent judgement. Suppose, however, that a community were to enjoin its members *always* to follow Kripke's procedure when applying natural kind terms. Communities apparently do not do this, any more than they insist on exclusive attention to immediate appearances in the application of terms, but they could seek to do so. Might it not be *possible* to impose rigid designation by rule and example, and does this not provide *in principle* a way of establishing taxonomic practices which escape the problems implied by finitism?

One difficulty with this is that rigid designation appears to be incapable of providing a complete and sufficient account of the use of natural kind terms. John Brown dies, and his corpse is no longer he; uranium becomes plutonium of its own accord, and may be transmuted into other elements if appropriately persuaded; mesons decay into pions. All natural kinds have instances which may cease to be what they are without any manifest spatio-temporal discontinuity. Continuity persists but is punctuated by a change of designation, generally inspired by some other form of empirical information. Death, transmutation, decay are the punctuation marks necessary to break and order the continuum.

Where natural kinds rather than proper nouns are the focus of interest the insufficiency of rigid designation is yet more immediately evident, for natural kinds occur in separate lumps and deposits, and the baptism of one lump establishes no connection of spatio-temporal continuity with any other. We may baptize a lump of South African metal as 'gold', but however carefully we keep tabs on it we shall not be led to Australian gold or Soviet gold by following the same-space relation over time. What else are we to resort to here, to connect one lump of gold with another separate lump, other than empirical appearances and properties? The answer must surely be that if the connection is to be empirically based there is indeed nothing else. (Kripke and Putnam attempt to solve the problem by moving beyond empirical matters altogether. For them, to identify two things as of a natural kind is to assert their *essential* identity, their 'sameness of essence', something analogous to John Brown's soul in lying beyond what can be empirically scrutinized. But to follow this argument here would take us beyond our immediate concern with the empirical basis of classification into issues to be treated in the next chapter.)

Continuity cannot be the sole basis for the application of a natural kind term. But it may in any case be argued that continuity, like similarity, is always a matter for contingent judgement. Kripke himself explicitly declines to give an extended and systematic account of his conception of continuity. This represents sound judgement on his part. The continuity of a boundary in a given location over instants of time is not a matter of an identical boundary existing in an identical position in space – to insist on identities here would actually destroy the utility of Kripke's notion. But when we move to the notion of a 'similar' boundary in 'roughly the same' location at a later instant – as we must to allow for the mobility and mutability of 'the same object' over time – the problem of resemblance returns once more. People may make out spatio-temporal continuity in different ways just as they may make out resemblance in different ways. And the fact that people agree in identifying spatio-temporal continuity as when, for example, they continue to agree in their designations of John Brown, is merely the same kind of contingent fact of agreement which people also manifest in practice in their identification of similarities of appearance.

3.3 BELIEFS

In the opening of this chapter we noted that classification is largely a matter of applying the terms for classes and kinds received from the ancestors. Systems of classification are a part of the inherited culture. But the acquisition of such a system is never just that and nothing more: it is invariably at one and the same time the acquisition of a system of beliefs. The processes wherein a collective transmits its taxonomies and its knowledge are a single process. People receive authoritative indications of what there is in the world and what is known about what there is together.

We have seen how the world will tolerate innumerable alternative systems of classification indifferently, allowing different cultures to establish their own particular conventions as they mark the similarities and differences between things. But beliefs about the world are more than conventionally accepted ways of clustering things. They are substantive convictions about what the world is like. They engender expectation. On whatever basis they are accepted, they are accepted as repositories of past experience which should continue to prove compatible with or be confirmed by current experience. People tend to trust the ancestors, but they also tend to trust their eyes. What then is the relationship between these two things: what are the roles of tradition and experience in the maintenance of beliefs?

Many attempts to answer this question speak of the relationship as one of conflict: people must choose between placing their confidence in the idols of the tribe and trusting the indications of experience. Reference to the ancestral beliefs may tell us one thing, and experience may tell us another. Tradition and experience may clash, and a choice between them will then have to be made. Or at least tradition and experience may clash until tradition is adjusted into full conformity with experience, and every next observation serves as further confirmation of what is currently known.

The flaw in this way of addressing the issue is that experience does not 'tell us' anything. The formulations of the ancestors are not contradicted by verbal messages from experience: experience possesses no language and is unable to speak to us. But do we not notice associations in nature, between things and their properties, between one thing and another, and so on – and do not the associations thus observed serve to confirm and justify specific sets of beliefs? The answer is that the associations presented by things no more serve to establish a specific set of beliefs than the resemblances presented by things serve to establish a specific set of classes. As Durkheim might have put the point: feeling the associations is one thing; the statement of belief is another. Indeed the roles of tradition and experience in sustaining a body of knowledge are very much the same as their roles in sustaining a taxonomy, and the discussion in section 4.1 of the latter problem serves well as a model in considering the former. The previously developed finitist account of classification is readily extended into a finitist account of knowledge.

Finitism applied to beliefs
All societies sustain systems of accepted beliefs, bodies of knowledge taken by members as true, or reliable, or worth betting on. Different societies sustain different systems, but any given system tends to be taken, by those for whom it is authoritative, as a valid record of past experience. As such, it is a source of expectation for the future, on

the inductive basis that the future will be like the past, and the past is something it has already well described. Thus, ancestral knowledge will be brought to bear upon current experience: the indications of tradition and the indications of the senses will indeed confront each other. What will be the result?

The first thing to recognize is that whatever the content of the ancestral belief system may be, no *fundamental* problems will arise in reconciling it with experience. This is a simple consequence of the finitist account of classification. Beliefs are formulated using the terms with which experience is classified. The finitist account of the application of the terms is also an account of what beliefs relate to. And open-endedness in the one implies open-endedness in the other. The simplest way of setting out what follows from this is to reformulate the five basic tenets of finitism as postulates about beliefs and bodies of knowledge: this will clarify the link with the earlier discussion in section 3.1, and ensure the continuing relevance of its arguments and illustrations.

1 The future implications of beliefs are open-ended.

   If what terms refer to is open-ended, then what beliefs are about must also be open-ended. Just as there is a sense in which we do not know what our terms mean, in that same sense we do not know what we believe.

2 No statement of belief is ever indefeasibly true or false.

   If we do not know what our terms mean, if what they are going to be used to refer to is not predetermined, then we cannot know what future cases will come to count as tests of our beliefs, and hence whether they are true or false. Nor is it a matter of mere lack of knowledge, as if a belief could be true but cannot be known to be true. Future cases are not merely unknown; they are not yet decided, and could be decided in the future in any number of ways, depending on the contingent judgements we have yet to make. Hence truth or falsity cannot be treated as inherent properties of beliefs. The terms should not be used to describe beliefs, and neither should any of the range of analogous terms which would attribute a fixed and definite epistemological characteristic to them.

   It might be claimed that a belief may sometimes be proved to be false simply by examining existing instances where it has already routinely been applied. But not even this is possible on the finitist view, because, formally speaking, past instances of use are just as problematic as future instances.

3 All existing exemplifications/confirmations/refutations of a statement of a belief are revisable.

Since any act of classification is revisable, since we are always at liberty to count it a mistake, what a belief was a belief about is likewise revisable. Whatever has counted as exemplifying, confirming or refuting the belief is always liable to re-evaluation and reclassification. It is always liable to be redescribed as 'not really the same' as the things which the belief relates to, or 'really the same after all' as those things. Any inductive support for beliefs must remain problematic accordingly, and any alleged 'falsification' of a belief must always count as less than that.

These three points, and perhaps especially the third, make it clear and evident that a belief may always be reconciled with experience at the formal level. Nothing stands in itself as an exception to a belief, since it may always be classified as something to which the belief does not refer. Given the indefinite complexity of experience and the lack of any metric for resemblance, good grounds may always be found for marking a difference between the exceptional item and those things which the belief 'really is about'.

For the sake of completeness it is worth reformulating the two final finitist tenets also, even though they are not required to carry the present argument. Both points reinforce the basic claim that a belief cannot be understood as a specific verbal formulation with a fixed and definite meaning.

4 Successive applications of a belief are not independent.

As a verbal formulation, a belief may be recognized as persisting over significant periods of time. But what is involved in accepting that formulation changes every time the belief is applied. Every new instance of what the belief is about will differ from earlier instances even though it is taken as 'the same' as those instances. Hence every instance wherein the belief is applied will change the precedents on the basis of which it is later applied: its application in one case will affect how it is applied in subsequent cases.

5 The applications of different beliefs are not independent of each other.

Every time a belief finds application the similarity relations of its constituent terms will change. But the same terms will appear in other beliefs. And changes in one set of similarity relations will in any event result in changes in the use of others, and hence affect the subsequent application of yet other beliefs. Thus, every judgement about the applicability of one belief will affect the future application of others: it will have a ripple effect through the entire system of classification and the entire network of belief.

Induction and the confirmation of beliefs
Just as the finitist account implies that nature is indifferent to how it is classified, so it implies that nature is indifferent to what is believed about it. Because it implies that nature is indifferent to how any preferred classification is applied and extended, so it implies indifference to how preferred beliefs are applied and extended. Formally speaking, anything goes. Far from telling us anything in particular, current experience can be described as consistent with any extant body of ancestral knowledge, even as offering inductive confirmation of it. Equally, it can be made out as refuting any specific body of knowledge or item thereof.

These claims have been vigorously advocated and exemplified in a series of case studies, by Harry Collins (1992). Collins denies the possibility of any confrontation of tradition and experience. Tradition may make claims, but experience cannot. Tradition may insist upon the existence of this or that form of order, or regularity, or association, in the world; experience itself gives no sign one way or the other. When human beings make sense of experience they read it in terms of existing knowledge without problem or resistance. Or else they are opposed by adherents of different traditions, or deviants or innovators in their own ranks, who invent alternatives to existing orthodoxy. Tradition is confronted by tradition, or by enemies within, but not by experience itself or the world itself. Reality has nothing to do with what is believed about it.[12]

Collins' method of choice is to focus upon significant scientific controversies in order to show how conflicting beliefs may be sustained about 'the same experience', and how reasonable human beings may actually take 'the same experience' as inductive confirmation for conflicting beliefs. The resulting studies are sufficiently well known to need no extended description here. Perhaps the best known of all concerns the controversy engendered by the claimed discovery of gravitational radiation in 1969 (Collins, 1975; 1992). The initial discovery claim stimulated a number of efforts at experimental 'replication' which involved a number of teams of physicists endeavouring to build 'gravity-wave-detectors'. The result was a set of similar-yet-different detection experiments, some of which gave positive and some negative results. In this situation the problem of whether gravity-waves had been detected was also the problem of which experiments were competent, and which detectors were correctly designed. If we knew how much gravitational radiation was around we could say which detectors were good, and which experiments competent. If we knew which experiments were competent, we could say how much gravitational radiation was around. Knowing neither, we can find out neither. Collins insists that this vicious circle always confronts the experimenter; it is the 'experimenter's regress', and it is only the breaking of the regress by whatever contingent means which allows scientific controversies to be resolved and scientific knowledge to accumulate.

The usual way of breaking the regress, according to Collins, is by reference to existing accepted belief: where scientists believe already in a phenomenon they recognize the competence of experiments which detect the phenomenon; where they disbelieve in the phenomenon they reject the experiment. On this basis those in the 1970s who disbelieved in the existence of significant fluxes of gravitational radiation could see their belief as confirmed by experience in the shape of data from 'competent' experiments, whilst those of the opposite view could likewise appeal to experience for inductive support.

Prior belief or disbelief in different kinds of phenomena is characteristic, according to Collins, of different forms of life. Hence controversies tend to be at their most intense and uncompromising when they occur between participants in different forms of life. It is this, rather than any technical incompetence or lapses in reasoning on the part of their opponents, which accounts for the intense criticism directed by many scientists against work on paranormal phenomena (Collins and Pinch, 1982; Collins, 1992, Ch. 5). Sceptical physicists may scrutinize unsuccessful attempts to evoke paranormal phenomena, and take them as matter-of-course confirmation of their already firm conviction that such phenomena are non-existent. Parapsychologists on the other hand may treat the identical states of affairs as yet more routine confirmation of the well-known fact that paranormal effects are inhibited by the presence of sceptical observers. Existing socially sustained convictions channel induction from experience along different paths in these two instances. Bodies of knowledge appropriate to two alternative forms of life are referred to and taken to be confirmed by experience. Perceived regularity in nature arises not from experience of nature but from the application of socially established conceptions of regularity in describing nature.

This is Collins' 'sociological resolution of the problem of induction' (1992, p. 145). Inductive confirmation requires that the confirming aspect of experience be 'the same' as something that has gone before. But experience does not tell us what is the same as what; it is a matter for us how similarity is weighed against difference. Hence coherent judgements of sameness, and consensually agreed relationships of confirmation and replication in science, only occur when the judgements are institutionalized and ordered as social conventions: the use of terms and the applications of beliefs must be *entrenched* in a way of life, with deviant usage sanctionable as a mistake in the use of language. When this is the case the familiar routines of confirming beliefs and replicating findings are possible as consensual operations, but 'It is not the regularity of the world that imposes itself on our senses but the regularity of our institutionalised beliefs that imposes itself upon the world. We adjust our minds until we perceive no fault in normality' (1992, p. 148). 'We perceive regularity and order because any perception of irregularity in an institutionalised rule is translated by ourselves and others as a fault in the perceiver or in some other part of the train of perception' (1992, p. 147).

Collins' detailed case studies of scientific controversy are among the best known of all investigations in the sociology of scientific knowledge, and rightly so. They figured prominently in that initial body of work which extended sociological study to the core activities of science, and analysed them as social processes. They helped to eradicate the view that sociological analysis was appropriate only for irrational action, and hence for understanding distortions and deviations in scientific reasoning; indeed they helped to turn that view upside down. Moreover, in uncompromisingly treating inductive inference as a social process they introduced a new and fruitful way of looking at induction. The vast existing literature on that subject had failed to do justice to its sociological dimension. It was largely the work of philosophers and psychologists who, for all their disagreements, were at one in their individualism. Whether it was a manifestation of universal reason or of personal idiosyncrasy, the process of induction was presumed to occur wholly within the confines of the individual mind. In Collins' work, on the other hand, inductive processes are displayed as the activities of a collective, and the more mundane, routinized and matter of course they are, the more they are revealed as conventionalized, institutionalized and entrenched as social practices.[13]

Nonetheless, Collins' approach conflicts with that taken in this book.

He offers a sociological account of natural order as a *replacement* for the individualistic accounts of philosophers and psychologists. Methodologically, it is an idealist account deliberately minimizing the direct connection of the individual and the physical environment. Order can be read on to experience but not read out of it. In contrast, where Collins insistently separates 'the natural' and 'the social' we see them as fused together; where he denies the relevance of the former, we insist upon it. For us, states of affairs in the physical environment have got to be taken into account in order to understand induction as a social process.

This way of thinking of induction is easy to illustrate using Collins' own examples. Collins notes that the negative experimental findings taken by some scientists to confirm the non-existence of extra-sensory perception are taken by others merely as evidence of its non-appearance on that occasion. Some students of paranormal phenomena participate in a form of life where the actuality of ESP is accepted, and these people will never infer its non-existence from any particular item of negative evidence. How these people induce is intelligible in terms of the form of life in which they participate. But is not positive evidence of ESP just as compatible with such a form of life? And, given that it is, why should they then speak of negative experimental findings rather than positive ones? Collins never poses this question. Our own view is that the physical environment is likely to be the relevant cause here. It was probably because of the state of the material world in the context of the experiment that ESP was considered absent; had the material world been in a relevantly different state, ESP may have been said to be present. This means that the account given of the experiment can only be understood by reference to the form of life being upheld by those offering the account *and* the state of the physical environment in the context of the experiment (where, of course, it is the unverbalized state that is being referred to, not any verbalization of the physical environment).

Pace Collins and his interesting idealist arguments, sociologists should be willing to acknowledge the existence and the causal relevance of the physical environment when they study the growth of knowledge. And having acknowledged this, they should acknowledge also the ability of individual human beings to monitor the physical environment and learn about it. Individual animals learn directly from experience. The psychologist's rat pushes the lever and looks to the arrival of a food pellet. If the pellet does not arrive, the rat gives every appearance of bafflement: it looks around, tries again, etc. The rat has learned to associate the lever movement with the arrival of food, and has developed an expectation that this association will continue in the future. The rat has successfully operated as an inductive learning machine: perhaps we should credit it with 'inductive reasoning'. In any event, it would be perverse to insist that what rats manage to accomplish in this context, human beings cannot hope to emulate; that what occurs in the brain of the rat is wholly distinct and discontinuous from what happens in the brain of a human being.[14]

It is more plausible to accept that the human brain, like the rat brain, may be profoundly affected by the reception of signals from the physical environment, and that these signals may directly engender not just perceptions and memories but associations and expectations as well. Consider a group of drinkers, thinking to pass the whisky but unknowingly passing the cyanide instead. Three or four corpses later, it is unlikely that any survivor would care to take a drink. In practically all social contexts, save only perhaps in a few London clubs, survivors would mark the association of drink and death, even if not in so many words. Whatever the cultural background or pattern of existing knowledge, they would rapidly infer that further consumption was likely to lead to further corpses. Such inferences would be manifested behaviourally in a refusal to continue to drink, very probably in a straightforward inability to drink further. Animal inductive propensities would have operated, and the drinkers would have found themselves gripped by the consequences. No doubt a sole survivor would be found in the same state of fear and trembling as a group of survivors, a state unlikely to be moderated by the thought that what remained in the bottle could be accounted the same as or different from that which had been imbibed already, as a matter of contingent judgement.

Of course, Collins is predominantly concerned with the confirmation of accepted beliefs, in communities where thought and action have become routinized and institutionalized. His central claim is that in this kind of context a sense of natural order and regularity derives entirely from culture and not at all from nature. Even this, however, must be called into question. Wesley Salmon recalls a colleague entering an aeroplane in the company of a child holding a balloon floating at the end of a string. Projecting from their existing knowledge, the passengers confidently expected the balloon to float toward the back of the plane as it accelerated into the air at take-off. Matter-of-course inductive reasoning on the basis of his (physicist's) knowledge led Salmon's colleague to expect the balloon to float forwards. He made a bet on the outcome, and had no difficulty with collection. Tradition confronted experience and lost. No persuasion was necessary: the indications of experience sufficed.

But in referring in this way to the role of experience and its power to induce changes in belief, are we not undermining the entire finitist point of view? Finitism claims that existing beliefs may always be defended, but we have cited a case where they were immediately set aside. Finitism claims that beliefs have no predetermined implications which experience may confirm or refute, but we have cited a case where just this appears to have occurred. In fact there is no real difficulty here, but the need to show that this is indeed the case allows us to conclude this chapter by setting out once more just what is and what is not implied by the finitist account.

First of all, although some emphasis has been given to the means by which existing beliefs and belief systems may be sustained and defended, finitism itself is in no way a theory of the stability of such beliefs. It treats stability and change in systems of belief indifferently as the results of contingently determined, pragmatically informed activities of reasonable human beings. If practices and judgements which reinforce existing beliefs are predominant that is an empirical finding which requires explanation. They are predominant, of course, and we shall go on to consider why later. In so far as the basic finitist approach itself is concerned, however, moves to sustain beliefs and moves to revise or reject them stand equally as viable possibilities, given the open-endedness of the correct application of terms and the associated judgements of sameness.

Secondly, it must be recognized that the reference to 'viable possibilities' here implies viability only in a very particular sense. The finitist claim is a *formal* one: it is compatible with specific individuals or collectives feeling compelled to believe A, or unable to accept B, just as we emphasized earlier that it is compatible with people feeling compelled to label objects as of one kind and not another. All finitism insists upon is that there is nothing in the meaning of a term, or its previous use, or the way it has previously been defined, which will serve to fix its future proper use, and hence the things which will confirm or disconfirm the beliefs in which it appears. Finitism denies logical compulsion, not necessarily psychological and/or social compulsion. It does not assert that individual human beings are free to pick and choose how they will use their terms, and what beliefs they will take to be valid. The formal defensibility of a belief must not be confused with its credibility. Credibility – what people actually believe – is an empirical matter. And the study of credibility suggests that it is most sensibly thought of as largely involuntary. We cannot in the normal way of things choose what we believe, any more than we can take a bottle of cyanide and toss a coin to decide whether or not we will take a drink.

In Salmon's example, particular people with existing systems of classes and kinds, and existing beliefs about these classes and kinds, paid direct attention to an element of experience. The routine matter-of-course operation of their perceptual and cognitive machinery led them to a conclusion at variance with their existing belief. Here was an object which did not do what they believed objects should do: it did not lag behind the motion of the plane as it accelerated forward. Nobody seems to have had difficulty in setting aside their existing well-entrenched belief about the mechanics of material objects. They were content to perceive 'a fault in normality'. They were content, moreover, even though the normality in question could have been defended and preserved with no great difficulty. Experience had made its effect, but it had not spoken; nor had it compelled the adoption of any particular belief or verbal formulation. Everybody took it that they had watched an object move forward as the plane took off; but it was open to anybody to claim that it was a hole in an object (i.e. the air inside the plane) which had moved forward, that existing knowledge was not wrong, but 'what it really implied' had been worked out with insufficient care. Nonetheless, everybody had been surprised, and the surprise derived from expectation which derived in turn from existing knowledge.

For particular human beings in particular situations direct attention to their physical environment may dispose them toward or away from particular convictions. There is a direct link between individuals and the world they live in; individuals do learn partly as a consequence of what passes along that link; how individuals learn from experience is a sensible focus of study. Sociological studies of learning and inductive inference serve not to replace but to complement existing individualistically oriented studies: the physical environment can have effects on the cognition of individuals no less than the social environment: if one makes a distinction between these two environments, then it is arbitrary and unjustified to recognize the role of the one and not the other. What the sociological approach may rightly stress is that individuals in different social contexts or with different histories may read what is apparently 'the same experience' in different ways; and that, even if they do not, the remote consequences of that experience for what they believe may be radically divergent yet equally defensible. New scientific facts feature amongst those remote consequences.

CHAPTER 4

Beyond Experience

In the previous chapter we considered how classifications might be applied, extended and modified in response to the indications of experience. We saw that although experience does not determine its own correct description, many acts of classification are nonetheless carried out automatically and routinely, that people in any given culture generally share an intuitive awareness of the resemblances and associations in nature, and classify on the basis of this shared awareness most of the time. No culture, however, operates in this way all the time. Nor is this simply because intuitions of resemblance are occasionally tentative or ambiguous. More profoundly, it is because people invariably find cause to discount some perceptions and observations, or to reformulate or reinterpret the statements which record them, in order to preserve the coherence of their shared scheme of classification and the knowledge which finds expression in terms of it. Automatic responses to experience are always liable to be overridden for the sake of coherence, for without it the communication and information exchange essential for the very existence of the culture would be impossible.

The linguistic practice which plays a crucial role in maintaining coherence is that of contrasting the 'real' and the 'unreal', or 'reality' and 'appearance'. Using a distinction of this kind, people are able to evaluate perceptions and observations, to select some as 'good data' indicative of 'how things really are', and to reject or reformulate the rest as in some way flawed or misleading. A *realist strategy* can be employed to assimilate new experience to existing knowledge, and to modify existing knowledge in the light of new experience, without loss of coherence.

Realist strategies in science

Some scientists are realists by philosophical inclination. More are realists in the sense that they are moved by explicitly realist forms of argument and persuasion. But all scientists are realists in the sense that they use realist strategies, as indeed are all human beings in all cultures. A number of examples of the use of realist strategies have already been considered in Chapter 1. They show how perceptions and observations are interpreted as real (of the world) or artefactual (of the perceiving organ, instrument or apparatus), in ways which sustain a single coherent account of what exists in the world.

Realist strategies come into play from the moment that an individual scientist makes an observation. Such an observation is invariably in truth a series of observations. A single instrument will be set up to make a series of readings, say of a specific heat or a melting-point. The readings are simply presumed to be imperfect observations of 'the' melting-point. Hence it is presumed that the readings should be the same, and that wildly variant readings may be discounted as artefactual. What most readings indicate is a good estimate of the 'real' melting-point; what single variant readings indicate is artefactual. The same strategy is likely to be repeated when the individual observation produced as above is added to the pool of observations in the possession of the collective. What many eyes or many instruments agree on will be an observation of what is really there; what just one eye or one instrument observes, in contradiction to others, will be artefactual. Or at least this will be the usual realist strategy. There will be cases where even shared perceptions and agreed observations are identified as artefactual: such are 'optical illusions', observable phenomena which are drained of ontological significance by the very way that they are described.[1]

In everyday life we take it for granted that we perceive the appearances of real material objects with distinct properties and propensities, and an independent existence in space and time. Indeed perhaps it is misleading to speak of anything being taken for granted here, as though we are aware of some assumption which we are making: perhaps it is better simply to say that we perceive a world constituted of the objects we believe to exist within it (Gibson, 1979; Harré, 1986). Just the same is the case in science. Scientists take much of their data as being about objects and entities which they believe to exist 'out there' in the world, and they use this assumption to discipline and reconstruct their data just as is commonly done in everyday life. But science has its own distinctive objects, objects not manifest in appearances and perceptions, theoretical entities like forces and fields, atoms and molecules, invisible and intangible yet nonetheless responsible, so scientists conjecture, for entire classes of observations and experimental findings.

Millikan's oil-drop experiment, discussed in Chapter 2, would now be routinely thought of as a measurement of a property of one of the basic theoretical objects of physics, the electron. Millikan's realist strategy, was to investigate not electrons, but 'the' electron and 'its' charge. That strategy is now institutionalized as a part of physics, to the extent that we even forget that an electron is a theoretical object, and treat it very much as a normal material object: we take the reality of the electron for granted. As scientists have gradually reorganized their actions around the premise that it is 'really there', so it has gradually acquired the status of a real object. Before this reorientation of scientific activity there was only 'Millikan's theory'.

This kind of transition is most clearly visible when realist strategies conflict. Millikan's experiment is indeed just such a case. Andrew Pickering has documented another, in the more recent history of particle physics (1981). In the early 1970s a large number of new elementary particles were reported, and much energy was put into developing a theoretical account of them. Eventually, two competing accounts emerged, both of which treated the new particles as combinations of yet more elementary particles – quarks: quarks were the relevant theoretical objects in both accounts, which differed in postulating the existence of different kinds of quarks with different properties. Thus 'the charm theory' and 'the colour theory' became recognized as alternative accounts of the new particles. By mid-1977, however, physicists had become almost universally agreed in their preference for 'charm'. 'Charm' and 'colour' remained viable competing realist strategies, but the pragmatic value of the former proved much greater in the current state of the discipline, so that it was widely used and accepted. Many disparate groups of physicists found 'charm' relevant to their particular interests and pragmatic concerns: not so with 'colour'. As a consequence 'colour' remained a theory – so-and-so's theory, as it were. But 'charm' became part of the institutionalized discourse and practice of the entire particle physics community, and ceased to be associated only with particular persons or factions. The charmed quark took on the same status as the electron. As Pickering himself put it: 'charm became entrenched in the practice of several sub-cultures of theoretical HEP, and in the process became real' (1984, p. 144).

Realism in science is considered in great detail in the literature of philosophy, which remains the indispensable resource for anyone seeking insight into the topic. But philosophers do not in general treat realism as a feature of the behaviour of scientists, and there is a great deal to be gained from treating it in this way. By identifying realist strategies as implicit in the collective activity of scientists we make their realisms into phenomena of sociological interest. It becomes apparent that their strategies are selected and employed to further the pragmatic concerns of the collective, and especially to sustain coherence, consistency and simplicity in their shared culture. In no way does this imply any criticism of science, but it does suggest that realisms in science are particular instances of something common to all forms of culture and implicit in all forms of practice. The presumed reality of ghosts and spirits organizes life in many tribal cultures. The presumed reality of caloric and phlogiston is used to organize life in scientific cultures. Until a case is made to the contrary, current realisms in science should be seen as analogous to these alternatives, and current theoretical entities should be set on a par with those encountered in the very different forms of culture sustained by aliens, ancestors and deviants. Scientists are distinctive in the theoretical objects they currently assume to be 'really there'. But in their sense that such objects are there, and in their use of the techniques and devices of the realist mode of speech to sustain that sense, they are typical of human beings in all cultures.

Realism and reification

Realisms vary, but the realist mode of speech is ubiquitous. People everywhere use it to sustain their sense of what is real. Social scientists interested in the process often refer to it as *reification*, and for the most part the term is used in a strongly derogatory way. To reify, it is alleged, is to lose sensitivity to the rich complexity of appearances, and to cease to be fully open to experience; it is to ignore processes of change and transformation, and to impose a hypostatized representation on a cosmos in flux; it is to set limits on the extent of our awareness and to impoverish our imagination; it puts groundless constraints upon what we believe to be possible and thereby upon what we will attempt to bring about. Thus, reification is attacked not just on technical grounds but on overtly political grounds as well. It is characterized as a reactionary ideological device designed to create spurious restrictions on our conceptions of political possibility, to exaggerate the stability of the status quo, and to dampen enthusiasm for social change.

A number of separate points need to be disentangled in response to this. First of all, such criticism suggests that reifications could be shaken off and dispensed with if we were of a mind to do so, which is surely to underestimate the extent to which culture and cognition generally are actually constituted by, and dependent for their very existence upon, reifications (Schutz, 1964; Thomason, 1982). What would it be like to live in a culture devoid of reifications, responding directly to the rich complexity of appearance, without simplification, without selectivity? It is hard to imagine. Perception itself seems to be pre-organized for selectivity and simplification. We are unable to perceive the world as a sequence of pure appearances, let alone to think of it in these terms. The realist strategies of a culture represent an extension and particularization of 'strategies' designed into the perceptual and cognitive process itself. There would appear to be no escaping a realist orientation to the world we live in and the ubiquitous conventions of the realist mode of speech. The realist strategy and the reifications it generates are indispensible, and accordingly stand beyond the kinds of stricture which would warn us to avoid their use or discard them in favour of something else.

When we turn from the mere existence of reification to some of the specific forms which it takes, matters become more complicated. The common-sense flexibility with which we deal with material objects is not the norm for all products of reifying strategies. Theoretical objects, for example, are typically treated not as variations on a theme, as books or animals or rocks are treated, but as *essentially identical* entities. Every electron is held to be identical in nature to every other. The charge of the electron is one of its essential properties, and as such is manifested identically in every particle. Whatever experiment and observation may say, it is legitimate to speak of *the* charge of *the* electron, just as we speak of *the* ratio of radius to circumference of *the* circle, and to take for granted the existence of that essential identity in nature. The *essentialism* apparent here is more than mere reification, and is more than what is necessary to the basic task of perceiving the world and developing a shared understanding of how it is structured. Identical electrons are as helpful in the culture of physics, as the constant is helpful in the culture of mathematics, but they are not necessary for the existence of the culture of physics.

Moreover, it is but a small step from essentialism to sacralization, and there is much to be said against the treatment of essences as sacred objects. When essences are held to be obviously and self-evidently there, and any suggestion that they are conjectural, conventional or constructed is dismissed out of hand; when they are proclaimed as fundamental unchanging constituents of the cosmos; when to question them is a moral failing, and to disregard them a sacrilege; then scientists may indeed merit criticism for undue authoritarianism and serious lapses from their self-proclaimed open-mindedness. And scientists do indeed sometimes proceed in just this way, defending their key theoretical entities as tribal societies defend their gods.

Should we then oppose any tendency to sacralization of the theoretical entities of science, whilst retaining an appreciation of the necessity and positive utility of the realist mode of speech? There is much to be said in favour of this, including the fact that the realist mode of speech serves not only to construct reifications and accounts of sacred objects, but to *deconstruct* them as well. One way of exposing the constructed character of a given version of reality is to criticize it by explicit reference to another. In *How the Laws of Physics Lie* (1983), Nancy Cartwright rejects cosmologies based on quantum mechanics, relativity theory and other formulations of grand theory in modern physics on the straightforward ground that these theories are false: they do not correspond to reality. Why is Cartwright of the opinion that this is the case? Because she is impressed by the 'lower-level' phenomenal theories of physics, and convinced of their truth. The phenomenal theories are inconsistent with the more general theories, and the latter are only kept in being by the concerted effort of physicists themselves, who employ, she says, a range of *ad hoc* devices and *ex post facto* rationalizations to conceal and compensate for their formal deficiencies.

Cartwright criticizes fundamental physical theory as not in correspondence with reality. But the leverage for her criticism is obtained not from reality itself directly, but from other physical theories presumed to be in correspondence with it. The overall argument for accepting the one set of theories and not the other constitutes a specific realist strategy. This is not to criticize the argument, which is reminiscent of the arguments scientists themselves use when they recommend the acceptance of one specific scientific theory over and against another. Nonetheless it is necessary to enquire into the standing of the 'phenomenal' theories which Cartwright relies upon in her evaluation, and to question whether the relationship of these theories to 'reality' is any less problematic than that of the fundamental theories which she calls into question. In her book Cartwright is content simply to record her own opinion that many 'phenomenal' physical theories are indeed true.

Cartwright's criticism of a particular set of realist strategies is at the same time a reaffirmation of realism. Naturally, many critics would wish to go much further than this. Indeed, for those of a strongly anti-realist persuasion, positivists, empiricists and idealists of other kinds, there may seem little need to make fine distinctions between realist modes of speech, reification, essentialism and sacralization. These merely mark, so it may be said, moves further and further away from the flux of appearances along a road which should never have been travelled in the first place: any world of objects must be rejected as a mere improvization upon appearances, any world of essences as a hypostatization of a cosmos in flux. Such criticism is in order if appearances are 'really there' to be improvised upon, or if the cosmos 'really' is fluid and protean. But claims of this kind are no easier to justify than their opposites. A discourse on appearances stands no closer to what we perceive than a discourse on objects: a discourse on the transience of all things is just as problematic as one on their stability and permanence. Radical anti-realists are liable accordingly to find themselves employing a realist strategy, and drawing upon the copious resources of the realist mode of speech in just the way they find objectionable in those they criticize.

Sociologists opposed to realism seem to have recognized this. Thus, those who 'deconstruct' scientific discourse and show how its theoretical entities are human inventions, rather than revelations of 'what is really there', generally accept that their deconstructions are themselves constructions, invented accounts of their own rather than revelations of 'what a theoretical entity really is'. In effect they accept that deconstruction itself amounts to a realist strategy, sustained by appropriate use of a realist mode of speech. But this does not amount to a problem for them. Their major aim appears to be to weaken the authority of accounts of what is real, and they have no need of any particular authority for their own accounts. They can allow that all speech, including their own, is a creature of a day. Nonetheless, this seems tantamount to an acceptance that it is not realism as such, or the use of the realist mode of speech, which merits criticism, but rather a particular attitude to realist accounts.[2]

The realistic discourse of modern science may be desacralized by a project of radical deconstruction, which nonetheless is itself obliged to make use of the realist mode of speech. Alternatively, a similar effect may be obtained by Cartwright's method of using a commitment to one set of theories to underline the 'invented' character of another set. But, as well as these, there is a third much more widely used way to avoid making a fetish of essences, and keep the authority of theory within bounds. What we generally do in everyday life is to contrast theory and indeed speech generally with reality 'itself', unverbalized reality, whatever is 'out there' independent of perception, thought and word. A naive common-sense realism of this kind has much to recommend it. On the one hand, in acknowledging a single external reality it justifies the imposition of coherence and consistency upon appearances and observation reports, which must be made intelligible as 'of' the same underlying states of affairs. But on the other hand, by refusing to conflate external reality with anything that is said of it, by keeping it beyond immediate appearance and hence beyond speech, it ensures that theories and descriptions remain visible as contingent human accomplishments liable to future revision or rejection. It reminds us of the corrigibility of existing notions, without asserting the superiority of any others. And it allows that future changes in what we believe may arise, not just in the light of verbal formulations of data or abstract theoretical arguments, but also because of causal inputs from the physical environment. As it happens, this naive form of realist strategy has sufficed to negotiate most of the major changes of objects and essences which make up the history of the natural sciences.

A naive realism of this kind is often dismissed out of hand on the grounds that the 'reality' it insists upon, the 'external' reality independent of the mind and its inventions, of society and its institutions, is a reality which cannot be described or specified, studied or examined. But this criticism is beside the point. It is in the way that they make reference to an external world beyond speech that people everywhere reveal their mastery of the realist mode of speaking, a mastery which keeps them safe from domination by the objects and essences which they themselves create. Naive realism represents a very general realist strategy which operates with marvellous efficiency in the assimilation of new experience; it is put to use to maintain, extend, revise and reject existing knowledge, and thereby serves both as that which sustains the status quo and that which changes it.[3]

4.2 THEORIES

A number of ways of studying realist strategies are available to social scientists. The literature of ethnomethodology is a particularly valuable resource, and its fount and origin, Harold Garfinkel's *Studies in Ethnomethodology* (1967), remains itself essential reading. This is a book with its own peculiar concerns, and it would be false to say that it is 'about' the issues which are engaging us here. Nonetheless, there is much to be learned from reading Garfinkel's discussion of 'documentary method' as substantially an account of the realist mode of speech.[4]

Garfinkel was initially interested in how it is possible to produce an adequate description of social and cultural events, and he suggested that 'the documentary method of interpretation' was necessarily involved in this kind of task. Following Mannheim (1952), Garfinkel characterized documentary method as a search for 'an identical homologous pattern underlying a vast variety of totally different realisations of meaning'. But this search is not intelligible as a search for some independent external entity. Rather, documentary method consists in taking 'actual appearance as "the document of", as "pointing to", as "standing on behalf of" a presupposed underlying pattern. Not only is the underlying pattern derived from its individual documentary evidences, [they] . . . in their turn are interpreted on the basis of "what is known" about the underlying pattern. Each is used to elaborate the other' (1967, p. 78).

Among the most memorable of Garfinkel's illustrations of these processes is the famous 'counselling experiment'. In this experiment, subjects were invited to address a series of questions to a 'counsellor' as a means of eliciting advice on a personal problem. The subjects were informed that the counselling session was at the same time part of a research project in ways of giving advice, and that every question would for this reason be answered in 'yes' or 'no' form. But they were actually offered a predetermined sequence of 'yeses' and 'nos' drawn up by use of a random number table. Subjects were thus encountering and responding to a 'meaningless' sequence of sounds, whilst initially presuming that a 'meaningful' pattern of interaction ('counselling') was occurring. What the subjects made of this was ascertained by asking them to break off from their unfolding interview after each question and answer, and to make tape-recorded comments on progress to that point.

Examination of the recordings showed that all subjects completed the minimum requested sequence of ten exchanges, and that all continued throughout to conceive of what was taking place as a manifestation of the presupposed pattern of 'counselling'. Presupposition, and not the pattern evident in the responses themselves, was evidently responsible for subjects' general sense of 'what was really happening'. The various 'yes' and 'no' answers were taken as expressions of a considered response to the problem brought by the subject to the 'counselling' session. The sense of any single answer was revised and reformulated retrospectively in the light of later answers, so that the overall sequence of answers remained intelligible as a coherent response to the problem. Concomitantly, 'the problem itself' was liable to be reformulated as the kind of problem to which the answers so far could be related as coherent and relevant 'advice'. When answers were found prima facie

inconsistent or incongruous, subjects were able to restore consistency
by developing appropriate conjectures about the 'counsellor', her state
of knowledge or lack of knowledge, or her ability to hear the question
asked as a question different from that intended by the subject: great
effort was expended on consistency-restoring activities when apparent
contradictions in answers were encountered. Although subjects did
occasionally become suspicious of the authenticity of the situation, and
might very well have come to regard it as a deception given enough time,
they were always able to make sense of it as a 'bona fide counselling'
situation.[5] They were able to do so because the 'answers', and indeed
the 'questions' and the 'situation', were, and remained after however
much elucidation, *vague,* and because such vagueness always allowed
a vast shared stock of taken-for-granted common-sense knowledge to be
invoked to elaborate new, more adequate versions of 'what the answers
really meant' (1967, pp. 89–94).

In a second, extended, example in *Studies in Ethnomethodology*
documentary method is employed in an altogether more consciously
'strategic' way. It concerns the case of Agnes, a young person living
as a female but whose otherwise unremarkably female anatomy included
possession of a fully developed penis and scrotum. Agnes had to refer
to a particularly unfavourable set of appearances, including prima facie
contradictory appearances, as documents of her standing as 'really a
woman'. The realist strategy necessary to her favoured mode of life
could not be sustained by routinized, habituated unreflective responses:
appearances were such that continual effort and social awareness were
essential if she were to pass as a woman. Indeed, one of the features
of the study is the way that it emphasizes the power of appearances,
the immediacy of physical and biological phenomena. The description
of the medical treatment which Agnes chose eventually to undergo
provides a good illustration: 'a castration operation was performed
. . . in which the penis and scrotum were skinned, the penis and testes
amputated, and the skin of the amputated penis used for a vagina while
labia were constructed from the skin of the scrotum' (1967, p. 121).
This is far more than gratuitous concern with detail. It emphasizes the
stubborn actuality of appearances, which appearances were precisely
the reason why Agnes had always to be so actively and obsessively
concerned with 'what she really was'.

For a number of years Agnes was confronted with the contrast
between the accident of appearances and the essence of her underlying
nature, both as an aspect of personal experience and a matter for public

consideration and acceptance. Hence, as an informant, Agnes could serve as a peculiarly rich source of insight into the use of the realist mode of speech, whilst at the same time standing as a typical instance of what its use involves. For what Agnes did with effort is just what everybody else does in the maintenance of their sexual status, albeit effortlessly. Nor did Agnes's high level of awareness of how realities have to be sustained lead her to reject those realities: Agnes recognized the reality of a world of natural males and natural females, just like anyone else; all that was distinctive in her case was the technical difficulty of sustaining the reality of that world as one in which she existed and always had existed as a natural female. Agnes was obliged to conceal misleading appearances from other people, so that they would not draw mistaken conclusions from them. In a search for an operation to remedy her anomalous anatomy, she was obliged to provide accounts of those misleading appearances to experts, so that, no longer concealable, they would not prove fatal to her objective. Her realist strategy proved successful in both contexts. She passed as a woman in everyday life. She obtained the operation needed to correct her anatomical irregularities and to make future passing that much less difficult.

It is an interesting exercise to develop the analogy between Agnes' use of everyday knowledge and a scientist's use of scientific knowledge. But how far can such an analogy be taken?[6] Garfinkel himself initially sought to show that documentary method was necessary both to participate in society and to make sense of what was going on therein. This made its use essential in sociological studies, and Garfinkel duly concluded *Studies in Ethnomethodology* (p. 288) with the suggestion that the reader review his earlier account of Agnes as itself an exercise in documentary method. On the other hand, there was no indication in this 1967 work that documentary method and the realist mode of speech were universal. On the contrary, they were contrasted with 'literal description', a form of description not possible for sociologists but available nonetheless to natural scientists. 'Between the methods of literal observation and the work of documentary interpretation the investigator can choose the former and achieve rigorous literal description of physical and biological properties of sociological events. This has been demonstrated on many occasions' (1967, p. 102). In subsequent years, however, Garfinkel and other ethnomethodologists have focused their attention explicitly on to the natural sciences, and many studies embodying their perspective have appeared. The distinction between documentary method and 'rigorous literal description' has been abandoned, and ethnomethods can now be counted among the available resources for the study of realist strategies in natural science.

In the natural sciences, the 'reality' which reports of appearances document is generally taken to be the reality described by a scientific theory or model, or constituted by the relevant theoretical entities. Any such report is liable to elaboration, modification or rejection, so that theory and observation remain mutually consistent. This is one sense in which accepted observation reports in science may be described as 'theory-laden'.[7] But care is necessary if this terminology is to be used. First of all, theory 'itself' does not condition observation reports. It is the judgements of the scientists who compare theory and observation which are potent here. Secondly, although observation reports are liable to modification in the light of their relationship with some given, accepted theory, they do not have to be modified. Observations and reports of observations may be unaffected by any given theory, including, notably, any given theory for which they are counting as a test or possible refutation. Finally, the notion of theory-ladenness is misleading if it is used to downgrade the importance of observation, and to give human ingenuity priority over sensory prompting in the process of knowledge generation. The general relationship of theory and observation is that of interaction. And even when theory is used to discipline observation, it is not a matter of cognition overriding perception or imagination governing experience. The accepted models and theories routinely employed by scientists have themselves a history, in the course of which they have been selected and adapted in the light of earlier observation. The interaction of theory and observation reports may be regarded as an interaction between past observations [condensed in theory], and present observations [reported as such]. What is denied when an observation report is rejected may be not the overriding importance of observation, but the priority of the here and now. Respect for theory may be a form of respect for the observational activity of the ancestors.

If observation is 'theory laden', theory is 'observation laden'. This point, apparently so obvious as to be trivial, is actually the key to understanding theory in sociological terms. Theory is documented by reported observations. Hence, as Garfinkel rightly emphasizes, theory will itself be perpetually changing as its documentation, observation, itself grows and changes. Moreover, that which is documented is known at any instant of time through its documentation. Theory will be known through the observation reports which are its documents. But as has been discussed extensively already, we cannot adequately specify what the terms used in these reports mean, or how they will be used on future occasions. It follows that even at a given instant of time we cannot hope to specify what a theory implies, and hence what precisely it is.

This may seem an extremely counter-intuitive result. Theory is commonly regarded as the logical core of scientific knowledge, that part above all others where strict definition and secure inference are established, where what assertions amount to is crystal clear even if their validity is not. Yet the result is entirely correct, and may readily be derived from practically any form of finitism.[8] A theory, presumably, must make reference either to invisible or visible entities: if to invisibles, then we cannot know its implications; if to visibles, then we cannot know its implications either, because of the standard finitist argument about the meaning of empirically instituted terms. Nor is the point without defenders among philosophers of science: even among those who have analysed in most detail 'the logical structure of scientific theories' there has long been a body of opinion which is in agreement with the claim.

Mary Hesse (1974) has suggested that models and theories in science are not sets of postulates from which deductions may be made to test against observations, but rather are resources for the metaphorical redescription of what is being observed or experimented upon. What becomes basic to a scientific theory is a body of discourse originally employed routinely to describe a familiar, well-understood set of situations. Scientists may use it to redescribe a different, less familiar, not well-understood set of situations, thus making it their 'theory' of the new situations. Thus, the range of audible phenomena may be redescribed on the basis of sound's being a wave; or the brain may be treated as a computer in efforts to account for phenomena related to memory or perception or inference; or a gas may be redescribed as a set of particles, with phenomena like pressure and temperature being understood as aspects of the motion of the particles.

Scientific theory is thus created by a shift of familiar discourse to new contexts, and phenomena in the new contexts become documents of the theory. The pattern and order in the familiar discourse become available to scientists as a framework which may create coherence out of the phenomena in the unfamiliar contexts. But the entire content of the familiar body of discourse is *never* carried over to the new area. Only a part shifts: we do not expect the brain to have a stabilized power supply, or a heap of particles slowly to build up at the bottom of the gas jar. And what discourse shifts and what does not as research proceeds cannot be predicted in advance. Nor does it settle and stabilize. As documents of the theory continue to be generated in the course of research, so the extent of the familiar body of discourse recognized as 'within the theory' is liable to vary.

May we nonetheless not characterize what the theory implies at any given point of time? Not so. Here is ⌐where the conception of theory as metaphor is required. When the familiar body of discourse is drawn upon to redescribe unfamiliar phenomena the redescription will have no institutionalized standing: it will be literally *absurd*, scarcely the property of an *implication* of a theory. Brains are not, literally, computers. Hence, we can dismiss any thought of the theory logically compelling a particular mode of use of itself, forcing scientists to apply it thus and so: that which is absurd cannot be logically compelled. Once more we must fall back on the idea that scientists' intuitions of resemblance and analogy are what move them to any particular form of metaphorical redescription. And once more we must recall that because of the problem of the resemblance relation, such redescription must be understood in terms of the contingent judgements of the scientists involved, and *not* as a consequence of what their theory 'really implies'. Moreover, following Black's (1962) analysis of the way that meanings on both sides of a metaphor are liable to interact and change in response to each other, we have to recognize that the familiar mode of discourse deployed in scientific theorizing will not be unaffected by that deployment. Our understanding of computers and our discourse upon them has changed through our attempting to treat brains as computers – perhaps it has changed even more than our understanding of brains themselves. Our understanding of particles has been affected by our treatment of gases as particulate. What a theory is becomes reformulated as it is used, so that even in an instant of its use it is not sensible to try to formulate 'what it really is'.

Practically any competent study in the history of science suffices to show that, whatever else, theory changes over time, so that references to theory X at different times are liable to be references to strikingly different things. Chemical atomism affords a standard example. Initially formulated, let us say, by John Dalton in 1803, the atomic theory of the elements has remained part of chemistry to the present, some would say paradigmatic for chemistry to the present. Yet if 'the theory' has indeed persisted over this period it certainly has not done so as an unchanged entity. Every statement made by Dalton in his celebrated initial statement is now regarded as false. 'The modern theory', if it could be formulated at all, would read more like a series of negations of Dalton's original postulates than a reassertion of them. Clearly 'the theory' has changed radically. It has changed so radically that it is worth asking what disposes us to speak of the continued persistence of 'the theory' at all. Perhaps a modification of Kripke's account of rigid designation, considered in Chapter 3 (3.2), would serve as some kind of an answer.

An even better example for our present purposes is 'Mendel's theory'. The alternative formulations of this theory in the literature are legion, and leave no doubt that the theory has changed over time. A recent textbook specifies Mendel's two laws: 'alleles segregate from one another during the formation of gametes' and 'alleles of different genes are assorted independently of one another during gamete formation' (Purves and Orians, 1987, p. 264 and p. 270). These authors offer an appropriately updated version of 'the laws', suitable for current teaching contexts; no doubt they are well aware that Mendel, in 1865, asserted nothing like this. He spoke neither of 'alleles' or 'genes', and he proposed no general genetic laws. The one law he did refer to was not a general law but 'the law valid for Pisum', and even that is not unproblematically identifiable with any specific verbal formulation in his text. But if 'Mendel's theory' has changed, what was it that changed? There is now such a detailed historical literature on Mendel's theory that this question can sensibly be addressed. We can focus on 'the theory' at a specific point in time and seek to characterize it. We can, for example, ask what 'the theory' amounted to in Mendel's own original presentation. This makes a particularly good choice, since the contents of Mendel's *Versuche über Pflanzen-Hybriden* have been extensively analysed by both historians and sociologists (Olby, 1979; Brannigan, 1981).

One thing which these studies clearly indicate is that, with very few exceptions, from the point of its 'rediscovery' at the beginning of this century until the present, successive generations of scientists have taken the *Versuche* to propound more or less 'the theory' which they themselves have accepted, so that the reformulations of Mendel's lecture serve almost as a recapitulation of the history of genetics itself. If the experimental results of the *Versuche* were documents of a formulated theory, then that formulation rapidly became itself a document. It was taken as itself standing for something else, some underlying pattern: 'what he really meant', or 'what he imperfectly formulated', where the 'reality' presumed was the 'reality' of the current science. A sin of this kind is routinely denounced as 'Whiggism' by historians.

Although it sidetracks us from our main theme, it is worth noting that whilst it makes for deplorable history this approach to Mendel does no harm at all to science. Consider all the generations of followers reinterpreting Mendel so that his theory expressed their reality. Clearly, none of these reifications of Mendel stifled original thought, since we have a whole series of them, not a single definitive version and then an orthodox church. As fast as one 'reality' was documented it was itself reckoned as a document, an imperfect representation of what the next version in the series would formulate. Such a series is often described as 'scientific progress'. In the achievement of that 'progress' the misinterpretations of 'Mendel's theory' were a part of that creative use of the realist mode of speech which engendered adaptation and elaboration of Mendel, movement beyond Mendel, 'advance upon' Mendel. And whilst there may be other strategies which could have achieved the same kind of effect, this one has the sociologically interesting feature that it serves its purpose whilst allowing no forgetting of the original 'discoverer of the theory', and without touching upon his honour.[9]

Scientists have had a vested interest in continuing to reinterpret Mendel, but what were they reinterpreting? What was Mendel's theory? If obliged, under coercion, to identify the best extant account of 'what Mendel was really saying', we should point to Robert Olby's brilliant (1979) review, wherein Mendel is taken to offer a theory designed to explain constancy and variability in plant hybrids. But the fundamental point which needs to be made is actually that *no* account will be unproblematically correct, in this case or any other.

Some slight insight into the particular difficulties of specifying 'what Mendel's theory really was' is obtainable by reference to an English translation of the *Versuche*. Mendel (1966) reports a series of experiments on the breeding of sweet peas. In the simplest of these, he takes two constant-breeding varieties of pea, differing in some specific characteristic, say colour or shape of seed, and breeds from them by cross-pollination. He observes how the relevant parental characteristics are distributed amongst plants in the next generation of hybrid pea and in subsequent hybrid generations, and shows that clear patterns are evident in these distributions. It is apparent that a characteristic of a parent may be absent in its hybrid progeny, yet manifest itself in a definite proportion in the next generation, produced by allowing the hybrid to self-pollinate. Mendel shows this to be consistent with specific assumptions about the invisible constitution of the hybrid plants, and of their seed and pollen cells (germ cells) and those of their parents.

Consider the case of a cross of wrinkle-seeded and smooth-seeded peas. Progeny from the cross were observed to be entirely wrinkle-seeded plants. However, these plants did not themselves breed true. Their self-pollinated progeny were only 75 per cent wrinkle-seeded: a 3:1 wrinkle:smooth ratio was observed. Moreover, when peas from the wrinkle-seeded progeny were themselves grown it was found that about one third produced all wrinkle-seeded plants, and two thirds produced 3:1 wrinkle:smooth. In modern terms, what is happening is roughly as shown in Figure 4.1.

On the modern view W and s are invisible elements (alleles) responsible for wrinkling and smoothness. To breed true a plant must have two like elements, WW or ss. Germ cells are produced by the separation of these pairs of elements, and new plants by the recombination of germ cells from the two parents, as in A above. If the element W is assumed to 'dominate' the element s, in the sense that any plant with one of each invisible elements looks wrinkle-seeded, then all the appearances in the breeding experiment are explicable in terms of the separation and recombination of the underlying elements.

Mendel's notation for this sequence, however, would have been a little different (see Figure 4.2). Notation of this kind is introduced by Mendel to identify plants by accessible empirical properties: their *characters*, or very slightly more problematically, their *traits*. The designations 'W' and 's' are assignable to plants on the basis of the shape of their seed. The designation 'Ws' is assignable where a plant has wrinkled seeds yet may pass on to later generations the smooth-seeded character: Ws is a trait visible by inspection of the seeds of the plant and the seeds of its progeny. To read this as an application of the modern 'Mendelian' theory, 'W', 's' and 'Ws' must be read as references not to traits, still less to characters, but to invisible elements or factors which transmit characters from one generation to another. And some account must then be given of why Mendel would, for example, write 'W' where he 'should have' written 'WW'. Olby is entirely convincing in holding that this form of reinterpretation is a strategic misrepresentation of a 'hypothesis of constant and independent characters which combine in all possible combinations' (1979, p. 67).

This cited hypothesis, however, cannot simply be read out of Mendel's lecture as 'what his theory really was'. What Olby has

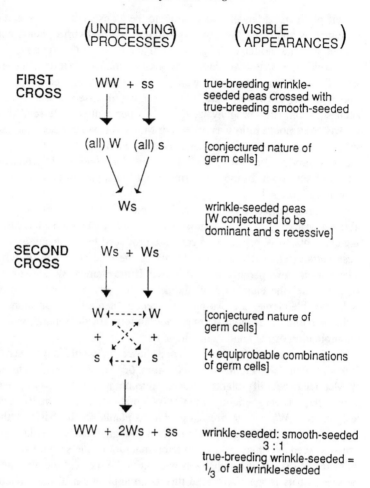

Figure 4.1

to do is to offer this interpretation as 'the best interpretation', given the text of the lecture, the circumstances of its time, the objectives of its author and the specific findings reported within it. Among the problems to be faced are: explicit references by Mendel to invisibles, such as the internal make-up of germ and pollen cells; references to elements achieving 'vitrifying union' in those cells; references to germ and pollen cells as of 'form W' or 'form s'; and even, in at least one

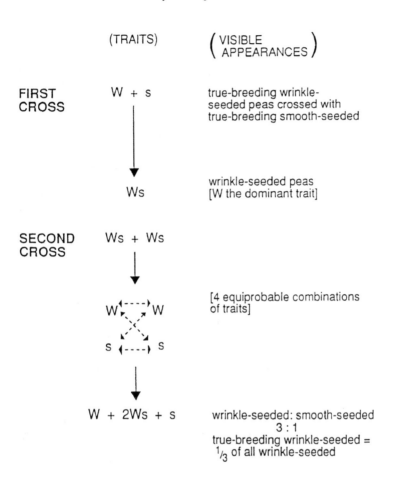

(TRAITS)   (VISIBLE APPEARANCES)

FIRST CROSS   W + s   true-breeding wrinkle-seeded peas crossed with true-breeding smooth-seeded

Ws   wrinkle-seeded peas [W the dominant trait]

SECOND CROSS   Ws + Ws

W — W   [4 equiprobable combinations of traits]

s — s

W + 2Ws + s   wrinkle-seeded : smooth-seeded
3 : 1
true-breeding wrinkle-seeded = $\frac{1}{3}$ of all wrinkle-seeded

Figure 4.2

place, a reference to a 'WW' kind of combination, the very combination central to the modern theory of segregating alleles and crucially 'absent' from Mendel's own account (cf. Olby, 1979, pp. 65–6, 61). A certain amount of hermeneutic art has to be employed before Olby is able to reconcile these matters with his preferred view. Which is not to say that, in so far as it is possible to identify 'Mendel's theory', Olby has not come as close as anyone, but only that it may be as well to recognise that in the last analysis it is impossible to produce a 'best' statement of 'the theory' and what it implies. After all, Mendel presumably considered that he had himself stated what his theory was, and what it implied. It is odd to imagine that another statement may somehow describe what is 'in' an initial statement, but not straightforwardly stated by that initial statement in the first place.[10] How is it that the best interpretation of a statement is not the statement itself?

4.3 PRACTICES

In the attempt to find 'Mendel's theory' the *Versuche* may have appeared as arcane, obscure and mysterious. In fact, it is pleasantly lucid and straightforward. It is mainly about what Mendel did with a large number of pea plants, of a series of observations he made upon them, and of the way he arranged his observations into patterns, using the routine procedures and manipulations of combinatorial mathematics. The words of the text are intelligible in relation to practices, either practices like cross-pollination which may be known already, independently of the paper, or practices like the mathematical manipulations incarnate, as it were, in the paper itself, and retrievable by an individual willing to 'go through' the paper in the appropriate way, 'doing the mathematics' as it is read. Even the most problematic and 'theoretical' terms in the text are intelligible in the immediate context of these practices: we understand what Mendel is doing with the terms nearly all the time. What else might be done with them, what they might 'mean' more generally, may remain obscure, but these problems no more threaten the intelligibility of Mendel's text than analogous problems in relation to the definite article, another term extensively used by Mendel, the 'general meaning' of which is less than fully clear.

In the *Versuche*, Mendel accounts for the distribution of specific characteristics in successive generations of pea plants. Nor is the accounting merely *ex post facto*. It allows prediction. Anyone taking similar plants, and breeding from them as Mendel did, is entitled to expect distributions of characteristics in the ratios Mendel describes. Thus, Mendel's work should be replicable, subject to the now familiar difficulties relating to 'same plants', 'same breeding', 'in the same way in the same conditions', etc. And indeed Mendel's claims have generally been acknowledged to be routinely replicable and to have been replicated.[11] But in the context of ongoing scientific research a paper like Mendel's is typically read with a mind to more than its immediate and routine 'implications'. If it is found important it is likely to be because it is found suggestive. Sweet peas have other characteristics besides those studied by Mendel: will they behave analogously? Cross-pollinations may be carried out using other plant species: will they give analogous lawlike results? What follows for other sexually reproducing species? What might be found by the microscopic study of the cells involved in reproduction? What might be the consequence of manipulations and modifications of the cells involved in reproduction?

The way to address questions of this kind is to do further work using Mendel's experimental practice and calculative procedure as a model. Experiments and observations on other characteristics, other species, animals rather than plants, cells rather than organisms, conceived in analogy with what Mendel did, are the route to ascertaining how far his achievement can be extended. In work of this kind, the question of what Mendel's more general terms mean and what his theories imply may arise as matters of practical concern. But even here it will be of little import what kind of an answer is given. For whether Mendel's more general verbal formulations imply, suggest, or merely hint at a particular line of development will be of scant scientific interest so long as the line of development is followed. The value of the paper for the research scientist will be that of a resource for research, a precedent and model for further practice. The use of terms in future investigations will be similar to, yet different from, the precise mode of use in Mendel's own. Hence any argument about what the terms 'really imply' and what 'Mendel's theory really is' will be impossible to resolve. Scientists will be able to make themselves out as applying or extending, correcting or transcending Mendel's theory as it suits them – say, according to whether they wish to bolster their work with the authority of Mendel, or find it more important to stress their own originality and creativity. Either way, Mendel's achievement will make its mark on future research.

Exemplars

Mendel's work on sweet peas is an instance of a scientific exemplar or paradigm. The terms are, of course, those of Thomas Kuhn. It was Kuhn's crucial insight that the fundamental units of scientific knowledge are not theories, nor even theories and associated observations, but solved problems. Problem solutions are the irreducible units of resource in scientific research: they are the models on the basis of which further problems are solved. Some of these problem solutions come to be recognized as holding a special promise for future research, and become accepted as authoritative bases for its practice in specific disciplines or specialities in science. These are exemplary achievements or paradigms. In them, theoretical discourse, practice and instrumentation are linked together and grasped in operation: they are understood in use in a way that they could not be understood by abstract consideration. And this understanding is the understanding of the collective for which the paradigm is authoritative, and thus constitutes both that which inspires research and that which informs its evaluation. To carry out research and to judge its results one must be able to do things with exemplars, not to make deductive inferences from theories. Hence, the difficulty of identifying the 'real meaning' of a theory, or even what precisely the theory is, has no adverse consequences for the actual practice of research.

What evidence is there to support the view that scientific knowledge is constituted largely of exemplary achievements and solved problems, and for treating these as 'prior to the various concepts, laws, theories and points of view that may be abstracted' from them (Kuhn, 1970, p. 11)? Kuhn himself cites the role originally performed by the classic texts in the teaching of science – Ptolemy's *Almagest*, Newton's *Principia*, and so forth – which he sees precisely as presentations of paradigms, exemplary achievements accompanied by specific illustrations of how to apply and extend them. He goes on to argue that modern science textbooks have just the same character, so that classical mechanics texts, for example, offer the student not laws and theories, but exemplary instances of problems solved by use of specific formulations of laws and theories, together with further problems, like the standard problems, for solution by the student. Direct examination shows that these standard problem solutions are not strictly deducible from laws and theories, whilst the reliable experience of generations of students is that their assimilation, by practice in their routine application and extension, is essential if the laws and theories of science are in a genuine sense to be learned. Students learn science not by learning rules for doing it, but by doing it, albeit in an attenuated form; they have to learn the vocabulary of science by exposure to concrete examples of how it functions in use: knowledge of nature and of language have to be acquired together.

Kuhn goes on to identify a direct relationship between the teaching of scientific knowledge and its use in research; between learning, applying and extending a body of knowledge. All these processes are aspects of the same process. The scientist is initially taught exemplars, and learns to apply them to closely analogous problems directly, in a simple form of modelling. This is what happens in examples classes, or laboratory practice, or in the exercises the student carries out in response to questions at the end of textbook chapters. In research, examples are similarly applied, this time to as yet unsolved problems, and the same modelling skills acquired during training are put to use. The processes are similar; the crucial difference is in the evaluation of the processes, where in research, unlike in learning, there is no correct answer to be found at the end of the textbook. In research, the scientist must hope that the analogy claimed between the solved and the unsolved problem will be an analogy found appropriate or attractive by fellow practitioners, that there will be agreement in practice amongst them all.

Training provides resources in the form of exemplars and solved problems, and illustrates their use in processes of modelling and analogy. Research is the further development of the same process. It renders the unknown in terms of the known, and hence allows calculations about the unknown to be made by analogy with calculations about the known. Once the connection between the known and the unknown is made, inductive inference can flow from the former to the latter directly, without passing through any 'general theory', and expectations can be developed about the unknown. Thus, knowledge can develop from case to case piecemeal in the process Kuhn describes as 'normal science'. The moves of normal science may be experienced as routine and matter of course, little more than the moves apprentice scientists make in their exercises. Or they may involve much less obvious analogies, analogies which only become generally evident after they have been made. Analogies of this kind are the key events in a tradition of research, the points at which new exemplars arise from existing ones. Kuhn's favourite example here is the exemplar of the point-mass pendulum (Figure 4.3.).

The motion of a mass concentrated at a single point, swinging freely at the end of a weightless thread of constant length, became a solved problem early in the history of mechanics. Galileo had obtained both that and the associated motion of a ball rolling down an inclined plane. He recognized that the key variables in both cases were height and speed; he knew how they were related; and he knew the specially simple conditions at the extreme points of motion, where speed became zero as height came to its maximum, and then speed reached its maximum value as the point of least height was passed through. To Galileo the two motions (parts a and b in the figure) were relevantly analogous.

As a solved problem the point-mass pendulum was recognized as

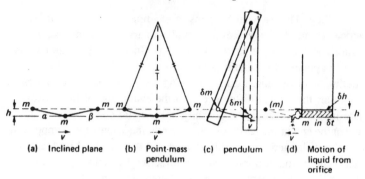

Figure 4.3

a promising basis for the computation of other motions: it became
a paradigm. A real pendulum made of a light string and a very
dense mass could be treated as an approximation of the point-mass
pendulum. More profoundly, Huygens used the paradigm to obtain the
motion of any real pendulum or swinging extended mass (such as, for
example, a ruler swinging from one end). The extended mass had to be
analysed into a large number of point-masses, so that the tendencies of
all the point-masses could be combined to yield information about the
behaviour of the whole (part c in the figure). This approach continued
in use in methods which employed the calculus to determine the centre
of gravity of a body, and then treated the body as a point-mass
located at that centre of gravity. These widely used methods exploited
mathematical techniques to extend the scope of paradigms based on the
idealization of the point-mass. Finally, Kuhn notes how Daniel Bernoulli
perceived the paradigm form in the unsolved problem of the flow of
water from a hole in a tank. Particles of water leaving the hole in an
instant of time must possess sufficient speed to carry them to a height
which exactly balances the loss of height of the water within the vessel
in the same time (part d in the figure).

The few brief pages in which Kuhn discusses the role of exemplars
are wonderfully profound and suggestive, and are the key to an adequate
understanding of his general discussion of scientific knowledge and
especially his conception of the nature of normal science (Kuhn,
1970, pp. 187–91). There is no material more relevant to the central
problems of the sociology of scientific knowledge, and it repays detailed
and careful study (Stegmüller, 1976; Barnes, 1982a). The deceptive
simplicity of Kuhn's ideas, their long availability, and the fact that crude

stereotypes of them have become widely disseminated and taken for granted, have all encouraged social scientists to rush rapidly through his work as a necessary but brief interlude on the way to the research front; but to do this is to risk reinventing the wheel. Fortunately, however, there have been some attempts to develop and elaborate upon Kuhn's account of the role of exemplars, and to add to the evidence which Kuhn originally offered to support it (Giere, 1988; Gooding, 1990). Kuhn's brief references to the problem solutions of classical mechanics have inspired Ronald Giere, who has made a detailed analysis of the contents of a number of elementary and intermediate textbooks in that field. Giere has no difficulty in showing that these books are indeed organized as a series of exemplars, that the exemplars are not derivable from the general formulations of laws and theories cited in them, or even necessarily consistent with those formulations, and that the 'theories' in question are not well-defined entities (Giere, 1988, Ch. 3).

Finitism and the use of exemplars

To describe the growth of scientific knowledge as a movement from one problem to the next on the basis of analogy and direct modelling is to offer a finitist account of the process.[12] Let us recall again the finitist account of classification. On this account instances of a class are designated, not by procedures fixed by definitions or rules, but by contingent social processes which take existing instances of the class as precedents. Every next instance is so designated on the basis of the analogy between it and existing instances, and the significance of that analogy for the agents doing the classifying. A class is its accepted instances at a given point of time: those instances are the existing resources for deciding what else belongs in the class, the available precedents for further acts of classification, the basis for further case-to-case development of the classification.

The relationship between a theory and its exemplary applications is like that between a class and its existing instances. How a theory is to be applied is not fixed by rules or definitions, or by deductions therefrom; it is decided by those using the theory, who take its existing applications as precedents and proceed from them on the basis of analogy. A theory is its exemplary applications at a given point of time, and that these are the resources for deciding what other applications will be made, the precedents for further problem-solving, the basis for further case-to-case development of 'the theory'. This is just Kuhn's account of research guided by exemplars.

Let us reformulate the central tenets of finitism with reference to exemplars:

1 The future applications of exemplars are open-ended.
2 No application/extension of an exemplar is ever indefeasibly correct.
3 The status of all existing exemplars is revisable.
4 Successive applications of the exemplars of a theory are not independent.
5 The applications of the exemplars of different theories are not independent.

These points offer a convenient starting point for further reflection on the use of exemplars, and if we wish to continue to engage with work which speaks of theories and their implications, they offer useful guidance as to how far speech of this kind is pragmatically acceptable. As we have explained already, theories cannot be identified as distinct entities which persist unchanged over time: they are not 'well defined', and it is not sensible to attempt to specify 'what they really are' (3, 4, 5 above). What theories are applicable to is indeterminate and yet to be decided, and since their scope is indeterminate it is inappropriate to consider whether they are true or false (1, 2 and 3). The fate of theories is bound up with the fate of other theories, and decisions taken concerning the use of one theory must be expected to have remote consequences (5). Since the cluster of exemplars making up a theory may be revised at any time, the specific cluster recognized at that time is a conventional ordering (3).

But what is it, in that case, which sustains the sense that a given conventional clustering of exemplars is one thing – a set of applications of an underlying theory. The way to explore this problem is to begin with the most radically naturalistic hypothesis, and suggest that this sense of underlying unity is entirely an accident of history. The process wherein the practical use of one exemplar leads to the production of another associates specific exemplars together in a tradition of scientific practice. These loose associations may then be systematized and arranged to correlate with the institutional boundaries of disciplines and specialities, and references to such-and-such a theory may then become assertions of the extent of the intellectual property claimed by a field, or celebrations of shared achievement by its practitioners. Certainly the tendency of scientists to refer to their overall disciplinary resources as a single theory can give a false sense of the unity and coherence of a hotchpotch of tools and instruments: statistical theory is a case in point.

It must also be recognized, however, that precisely because of the manner of their historical development, examination of exemplars does often give rise to 'feelings of resemblances'. In his published lectures, Feynman (1966, Ch. 21–1) describes the study of different problem solutions in physics as the study of the characteristics of a single differential equation in different conditions; and there is indeed no denying that similar mathematical operations recur in exemplar after exemplar. This points to a pragmatic–technical rationale for grouping such exemplars together as instances of 'the same theory': they are easier to learn in that guise; mastery of one problem solution greatly facilitates mastery of the next. But Feynman's observation does not imply that different exemplars are deducible from a single theory (equation). Sneed (1979) has emphasized how the different physical constants which appear in different problem solutions serve to connect them together and make the data emerging from the use of one relevant to the use and appraisal of others. Thus, the velocity of light appears as a universal constant in many exemplars in physics, and the mass of the earth as a presumed constant in quite a few: these must each have a single value, and this requirement makes the development of the different exemplars in which they appear mutually interdependent and mutually constraining.[13] Again, this suggests a pragmatic–technical rationale for grouping exemplars together in specific ways, but again it does not imply the existence of a single theoretical structure lurking behind all the various exemplars. The cross-linkages identified by Sneed have, after all, not prevented radical historical changes in the ways that exemplars have been clustered. (Some of the discussion in section 3.2 serves to reinforce this point.)

To return to the matter of the application of exemplars: we have characterized this as a process of modelling, but what is involved in modelling? This is a vast and difficult subject, and even the extensive literature in the philosophy of science only manages to scratch the surface of it. It must suffice here merely to note two important ways in which modelling differs from classifying, and hence in which a finitist account of modelling requires an elaboration of our earlier basic account of finitism. First of all, modelling links many kinds of thing: it may relate two verbalized systems, or mathematical structures, or pictorial representations, or scientific instruments, or aspects of the material world, or it may relate any one of these to any other. And this means that although the relationship between the model and that which is modelled remains one of resemblance or analogy, the analogy need no longer be a perceptual analogy; it may be analogy through one-to-one correspondence of points, or through functional equivalence, or yet other kinds of relationship. Secondly, when modelling rather than perceiving is involved, opportunities for active intervention are particularly clearly apparent: scientists will be able not only to modify and adapt the model, but in many cases that which is modelled as well.

Aspects of whatever is providing the basis of a model may be eliminated by processes of abstraction and idealization, or elaborated and made more specific. It is possible to move to a world of massless strings, frictionless surfaces, and unresisting air, in order to develop exemplary methods of mathematical calculation in mechanics, and then to put the masses and the friction and the air resistance back into the initial exemplars, in order to extend their scope as models with practical relevance. And as with models, so with what is modelled, active intervention by the scientist may be possible and profitable: mechanical systems with thinner strings, smoother surfaces, and emptier spaces, may be constructed as appropriately simple realities to be modelled, and hence to be understood, predicted or explained. Or again, elements may be added to what is modelled: lines added to geometrical figures by construction may transform them into simple instances of existing exemplars; impurities introduced into unduly perfect crystals may transform them into well-behaved semiconductors; anomalous experiments on sunlight may be made exemplars of Newtonian orthodoxy by using a better prism, or using the prism better (Schaffer, 1989). Modelling, then, involves not just judgements of similarity; it involves manipulation of both model and what is modelled to construct similarity (Gooding, 1990) – or occasionally, especially in scientific controversy, to destroy similarity. But, as with judgement, so with manipulation; it is not fixed or defined by the model itself, by what is modelled, or by the initial relationship of the two: indeed, all of these are liable to change in unpredictable ways as modelling proceeds.

To identify modelling in science as contingent action may be obvious, but it is crucial. When it is overlooked, the result is typically a purely formal account of modelling, which fails to grasp its purposive and goal-oriented character, and hence how it comes to be recognized as successful or unsuccessful. There is no perfect model. To take A as a model of B raises the same issues as taking A as similar to B: the notion of perfect modelling is as problematic as that of perfect similarity (identity). Accordingly, a successful model is not identifiable by the 'correct' matching of model and what is modelled in some abstract, disembodied sense. A successful model is a pragmatic accomplishment, something which those who evaluate it take to serve their purposes. Success occurs at that point in the process of adjustment and adaptation when model (exemplar) and modelled (problem) are brought into a maximally effective relation for practical purposes. And this in turn, if it is correct, means that to understand modelling and the use of exemplars in science it is necessary to consider the practical purposes of the users. Thus, when Harry Collins studied the use of a working laser as a model for the design and construction of further instruments, the criterion of successful modelling was the generation of coherent radiation. Until the new instrument lased, modelling had failed. When lasing was achieved, modelling had succeeded. Many ways of conceptualizing the relationship between the original instrument and its 'replicas' were 'in principle' possible. Only one mattered in the research context itself, that which fitted with the goals and objectives of the constructors (Collins, 1992). In other contexts in science the goals and objectives which inform research, and make sense of the way that exemplars are used and developed, may be much more difficult to identify and describe. But it is difficult to see how any of these processes of modelling are to be understood, other than in the kind of pragmatic terms set out above.

It only remains to move the analogy between modelling and classifying into reverse. If modelling needs to be accounted for pragmatically by reference to practical purposes and goals, then so too does classification. When scientists employ exemplars it is important to ask 'what are they trying to do?'; but so it is also when they apply classifications. Indeed the question needs to be asked of all scientific actions, of observing and representing as well as experimenting and intervening: all these are contingent actions covered by the finitist account.[14]

CHAPTER 5

Sociological Projects

In the previous chapters we described a range of scientific activities, paying particular attention to the role of tradition, convention, agreement, and the social processes wherein such things are sustained or undermined. The intention was to emphasize what is otherwise liable to be overlooked or underestimated, so that the proper scope of sociological approaches in the study of science can be appreciated. Now, however, all this will be taken for granted. We shall simply assume that science is a routine instance of what social scientists study, and ask what that study involves.

The objective of the sociology of science is to describe scientific research as action, and to understand scientific knowledge as implicated in and produced by that action. Scientific research is what scientists collectively do; sociology is concerned with what people collectively do, with how and why they do it, and with what consequence. Such a project needs no extrinsic justification, but it does as it happens have considerable extrinsic value. Not least, it reveals the limitations of existing stereotypes of science, stereotypes which are relied upon as we orient ourselves to scientists and the claims they make upon us, but which are *not* derived from detailed familiarity with what scientists actually do. In accepting a responsibility to describe scientific knowledge and practice as it actually exists, sociologists have found themselves calling into question unduly idealized versions of science, and in that way have contributed to the development of a well-informed society.

But how does such a project differ from that of the historian? There need be no difference. Indeed, social scientists and historians of science have rightly recognized this, and tended less and less to concern themselves with demarcating strong boundaries between their fields. But it is nonetheless the case that different academic fields have different resources to bring to bear on the task of description. Social scientists will be disposed to describe science in terms of the particular models available in their tradition.

Curiously, when the sociological study of scientific knowledge initially developed about a quarter of a century ago, there was no obvious model for the field to build upon. Sociology had no accepted, well-exemplified way of describing technical activity, or even everyday knowledge-based activity. Accordingly, people tended to draw upon a range of diverse resources: a small literature on the sociology of knowledge which was actually mainly concerned with religious doctrines and political ideology, work on the professions, studies of socialization and cultural transmission, and, perhaps most valuable of all, anthropological studies of knowledge in tribal societies. Nonetheless, this pot-pourri, together with the general background of the field, did suffice to produce a shared if loosely drawn picture of what knowledge was 'normally' like.

In brief, knowledge in this normal version was the possession of the members of a culture or subculture, transmitted from generation to generation as a part of their tradition, and dependent for its credibility on their collective authority. Its use and application were not amenable to description in abstract logical terms, but had to be understood in relation to what people were doing in specific contexts for specific ends. And to understand its growth and change it was likewise necessary to attend to its context of use and the particular practical concerns of its users. This, then, was the kind of model sociologists had available when they turned to the systematic study of scientific knowledge, and an important question for them was of course whether or not it was an appropriate model in that context.

What they initially found was that the normal model was very strongly at variance with accepted thinking. Scientific knowledge was conventionally understood as the accumulated product of the actions of independent rational individuals. Yet it did nonetheless prove possible to assimilate scientific knowledge, and the processes whereby it was sustained, modified and transmitted, into the normal view, and to cast doubt upon the empirical adequacy of the previous individualistic account. Sociologists were able to recognize scientific research as analogous to cultural processes already described in their model. And they were able to identify and make use of those existing general accounts of scientific knowledge with which the model had affinity.

This, of course, is why the writings of Thomas Kuhn have been so very important in the sociology of scientific knowledge. His account of science could be read as an application of the normal model, indeed as an extraordinarily brilliant one, one which could actually serve as an exemplar in its own right, not just for further analysis of science, but for redescribing bodies of knowledge in other contexts. Kuhn (1970, 1977) identified scientific knowledge as the possession of a collective, incarnate in their tradition of research. He offered detailed insight into the nature of scientific training, and described how the transmission of knowledge and competence relied upon trust and the acceptance of authority. He pointed out the continuing importance of authority and collectively agreed judgement in the process of research itself. He demolished the standing of formal abstract representations of knowledge as accounts of its actual character. He firmly established the need to evaluate knowledge in a context of use through his discussion of exemplars, and thereby at the same time demonstrated the existence of a tacit component in all bodies of knowledge. And he argued that competing systems of scientific knowledge, tied in as they were to alternative practices, were formally incommensurable, which, as well as its relevance in philosophical debate, served to vindicate the sociological conjecture that the normal model would apply everywhere, to any body of knowledge or knowledge-based activities. Moreover, the fact that Kuhn was a respected historian of science, writing mainly to the problems of history and philosophy, merely increased the sociological significance of his work. For it stood as a convincing application of the normal model, whilst being by no means dominated by a sociological perspective or unduly dependent on the resources of the sociological tradition. Kuhn, as supporters and detractors alike have always agreed, possessed a fine command of technical detail in his work as a historian, and his advocacy of what was to all intents and purposes an extension of sociological approaches into the study of scientific research was all the more powerful for that.[1]

Theoretical description
The outcome of the extension of the normal model will be evident in the earlier chapters of this book, just as it would be in the rest of the literature in the sociology of scientific knowledge. A new description of science has emerged, widely recognized, and not just by social scientists, as preferable to the earlier individualistic account. Of course, the new description is no more a simple reflection of the reality of science than individualistic accounts were: this is indeed an implication of the new description. But there is no need to put a pessimistic gloss on this

fact: it should serve as a reminder, not of the limitations of description, but of one of the fundamental sources of its value. Any act of description activates the conventions of an existing vocabulary of description and thereby connects what is described with other things and other situations; it thus allows expectations, competences and experiences to be generalized and made more extensively applicable. Every act of description clears a pathway for the imagination to run along.

As far as the specific project of the sociology of scientific knowledge is concerned, to describe science according to the normal sociological model is to treat it as the same as knowledgeable activities in other social contexts; this allows expectations to flow back and forth from one context to another, and competences, procedures and representations to extend their range of application from one context to another.[2] It makes what is known of the role of cosmology in tribal societies, for example, available as a potential resource for an understanding of science; and of course the relationship is two way, and allows descriptions emergent from the study of science to be used to throw light on issues and problems elsewhere in the social sciences. Moreover, all this is made possible by matter-of-course description, in no way to be set apart from processes of description generally. There may be a case for speaking of 'theoretical description' here, but only as a reminder that description *always* involves specific cultural resources, and that the consequent extension of relations of sameness is *always* the conventional action of a particular collective. The term may perhaps also serve as an antidote to the idea that for a given state of affairs there is just one correct description: this odd notion, which invariably disintegrates on direct confrontation, nonetheless seems prone to creep into our thinking unconsciously, and does need to be guarded against. But if we are to refer to 'theoretical description' on these grounds, it should be in the knowledge that two words are being used when one will do.

Over the last twenty-five years, then, sociologists have redescribed scientific research as typical knowledge-based social activity. But they have not been content to do that and no more. In describing actions, institutions, social relations, groups and collectives, they have been forced to consider what these things are, and how it is that they are. In particular, they have had to confront the perennial problem of the relationship of the individual and individual action to the collective, whether as a group, or a social structure, or an institution, or a culture. Concern with this kind of problem has never been far from the surface in the literature of the sociology of science.

The first body of systematic work in the post-war sociology of science, that of Robert Merton (1973) and his school, serves well to illustrate this. As well as a description of science, it was a reflection on the problem of social order and the relationship of the individual and the collective. It addressed itself to the currently most promising efforts to solve these problems, applied them in its analyses of science, and emphasized the relevance of its findings to the evaluation of these efforts. In this Mertonian sociology, scientists were described as members of a moral community, acting in conformity with accepted norms of conduct, within an institutional framework where the honour advanced or withheld by scientific peers served as a crucial additional encouragement for them to act for the collective or institutional good rather than their own immediate individual benefit. Unfortunately, this fine work was mainly concerned to describe the systems of communication and reward in science, and was never systematically extended to describe the technical activities of research itself, or the detailed responses scientists made to each other's technical procedures and knowledge claims. There were hints of the promise of the approach in this context, but no more. This is a pity, because the vision of the individual human being implicit in this Mertonian sociology is perfectly compatible with the normal model, but is not currently shared by any of the other groups of sociologists currently working with it.

The trend in the sociology of science since Merton has been to emphasize more and more strongly the standing of individuals as active agents in order to make sense of their actions or explain their provenance. This is consistent with an analogous shift in the social sciences generally. How far this shift is itself a part of a much more broadly based pattern of social and cultural change is a matter for conjecture: that it has actually occurred is everywhere recognized.

In the sociology of science this changing conception of individuals and their relationship to the 'social context' could be expressed in two ways. One was to revert to aspects of the 'economic' conception of the individual to understand what scientists did. The other was to continue to stress the importance of culture, collective representations and traditions in science, but to reconceptualize them as facilitating resources for, rather than as constraints upon, individual actions.[3]

Understanding the social interactions of scientists in terms of their individual expediency has long been recognized as an important approach in the sociology of science, but until recently as one of limited scope (Ben-David, 1984). Some idea of what can happen when it is uncompromisingly extended and generalized may be gained from parts of Bruno Latour's book *Science in Action* (1987). Science, Latour says, is war; the object of war is to win; each and every scientific action is a move in a game where the objective is to win. The war, moreover, would seem to be Hobbesian war, a war of everyone against everyone; for science is chaos, according to Latour. Thus, if scientists make an alliance, it is an expedient linkage, oriented to 'winning', expendable immediately circumstances require it. If they accept a knowledge claim, it is accepted out of expediency, and will be set aside immediately circumstances require it.

An account of this kind immediately raises problems in understanding the apparent order and stability that does exist in the interactions between people. Indeed sociologists have long used Hobbes as a *diabolus* to raise all the problems of social order they wish their theories to account for: Merton's approach in the sociology of science, for example, belongs in a larger tradition of work which takes for granted these problems and the fact that they are insoluble in the framework of an individualist theory, where agents lack concern for one another or mutual susceptibility to one another. It is worth recalling that this tradition, which can be traced back all the way to Durkheim and his strategic arguments against the economists of his time, inspired one of the finest of all sociological studies of war, one which supported the Durkheimian conviction that what it needs is love (Stouffer, 1949).

Latour recognizes the interest and importance of these problems of order, and the 'actor-network theory' which he has proposed as a way of describing the entire range of scientific activity can be read at the same time as a response to them. Although many of the ideas and procedures of 'actor-network theory' are now widely employed in the sociology of science, it remains difficult to evaluate because of the diversity of the many elements which have been drawn into it. There is, however, a fundamental problem which has to be faced by any approach of this kind immediately it aspires to complete generality. To treat all the activities which generate and sustain knowledge as calculated, expedient 'political' actions is to treat them all as informed, *knowledgeable* actions. Hence it is inconsistent to assert that knowledge is only ever accepted on calculated and purely expedient grounds. This produces the same kind of circularity as the assertion that language is the outcome of extended negotiations on what words should be taken to mean.

It is important however that we take note of the virtues as well as the limitations of Latour's work, so let us praise it for its methodological significance. It encourages anyone disposed to consider science from the perspective of political economy to consider every single action without exception in that light, and to refuse to exempt any aspect of science whatsoever from that kind of scrutiny. This is indeed the properly consistent approach. The study of science has long been plagued by dualisms and a priori convictions about what is sacred and what profane. Latour's approach is insistently monistic, both in intention and execution.[4]

The other way of presenting the individual as an active agent in the context of the sociology of science is to characterize him or her as a participant in a 'form of life'. The term is Wittgenstein's, and its use here is testimony to the relevance of Wittgenstein's work, directly or indirectly, to the work of many sociologists. Those who have taken up the work of Thomas Kuhn have thereby linked themselves to Wittgenstein; so have those who have extended ethnomethods into sociology of science. Harry Collins, who makes the most frequent explicit references to forms of life in science, has used the work of the philosopher Peter Winch as a line of access to Wittgenstein's ideas. The finitist account of the use of scientific knowledge in this book is another version of the same position.

Kuhn describes science as so many paradigm-sharing communities. The individual scientist is a member of such a community, and his or her relationship to it is a matter of legitimate conjecture. Some writers have thought of it as comprising an active personal commitment to given beliefs or rules or methods; others have thought in terms of social constraint. To take the scientist to be a participant in a form of life is to insist instead on the relationship as facilitative. The individual, by virtue of a training, participates in a tradition, and has available what it provides. In making use of what is available, in whatever way, he or she contributes to the continuation of the tradition. The tradition does not 'make' participants use it in this way or that; rather, in using it in this way or that they constitute the tradition. This conception of participation needs to be extended to all the elements of tradition, even the most apparently constraining ones: laws, rules, norms, meanings, definitions. Participants are those who share the elements of the tradition, not those who deploy them in one way rather than another, still less those constrained or determined by them. Wittgenstein speaks of the agreement in the practice of the participants in a form of life as agreement not in their opinions, but in the language they use (1968, Section 241).

But if we take scientists to be participants in a form of life, and scientific research to be the active articulation of a received tradition in a series of contingent actions, how can we do justice to the manifestly coherent and ordered character of scientific work? If individuals participate in a tradition but develop its resources every which way, simply as whim and fancy dictate, will not the tradition itself rapidly disintegrate? Whereas if individuals develop such resources coherently, in concert, so that tradition is reconstituted as it is used, is not our account of what is going on radically incomplete?

Harry Collins faces this problem in *Changing Order* (1992), where various scientific controversies are understood as confrontations of different forms of (scientific) life. His solution requires an additional postulate: the given form of life in which the routine ways of using tradition are entrenched is what nearly everybody likes, and the larger and more pervasive the potential changes in the entrenched, routinized ways, the fewer the people who will like them (1992, p. 133). Tradition itself does not constrain: people may take it on just as they will. But the routine ways suit most of the people most of the time, and in being generally followed for that reason ensure the continuation of the tradition; which is not to say that some individuals will not diverge from the flow of routine, nor that others will not follow them at times, to establish new routines and new traditions. This additional postulate of Collins' represents an attempt to *explain* the particular paths along which scientific practices develop, and why they are, as it happens, predominantly orderly, stable and routinized.[5]

5.2 EXPLANATION

Participation in a form of life does not involve conformity to fixed patterns. Nor does it even require a willingness to conform to fixed patterns most of the time. The specification of given routines or patterns will not suffice as an account of a form of life. We recognize a continuing form of life where people *remain coordinated* in their use of shared knowledge/culture, in their practice, even as and if the practice itself develops and changes. A form of life has much in common with an individual life as described by Kripke (section 3.2 above): it is to be identified more by its continuity than by its characteristics. Thus, of any particular action within a form of life, it is always legitimate to ask for further insight into why it was that action and not another which was performed in that setting on that occasion.

Whilst a 'complete' account of the provenance of any given action is not to be looked for, the particular actions wherein scientific knowledge is applied and extended need to be understood in as much detail as possible. As well as being intrinsic to the sociological project, this is a task of considerable extrinsic importance, for it is these actions which give scientific knowledge its far-reaching practical significance. What scientists do in the laboratory and what the rest of us do outside it are connected by an accepted judgement that this or that within is *the same as* this or that without. On the basis of such judgements we adjust the state of our physical environment, adding appropriate substances and removing others; the state of our bodies, adding and removing in the same way; and the state of our society, adding and removing people appropriately. Sometimes we decide that something in the laboratory was after all not really the same as something outside, and we alter our methods of waste disposal, or our register of drugs, or our prison population, accordingly. The consequence of this kind of activity is hard to exaggerate.[6]

If we wish to account for applications of scientific knowledge as particular actions, then the most straightforward way to proceed is by use of the language of cause and effect. A concept is applied, or an exemplar is extended, in one particular way rather than another: A is taken as the same as B, when it may equally well have been taken as different from B. The more we are able to identify causes, relevant necessary conditions, germane to its use in the one way and not another, the closer we have moved toward a satisfactory explanation. This approach, however, is almost everywhere found unacceptable in the context of the sociology of scientific knowledge, and the present authors are practically the only members of the field to make use of it.

Some opponents of causal explanations of actions assume that causation means *external* determination, and hence treats individual human beings as no more than empty puppets responding to outside prompting. But this assumption is surely incorrect: causal explanations are not bound to cite external influences only; they may legitimately refer to the internal state of the system whose behaviour is at issue, as when, for example, the operation of an internal combustion engine is explained in causal terms. Even when 'internal' causation is recognized, however, causal explanation is still liable to be criticized as logically inconsistent with belief in human autonomy and the notion of the individual as an active agent, capable of judgement, decision and free action. The response here must be to ask where precisely the alleged logical inconsistency lies, and to suggest that when a detailed inspection is made, no such inconsistency will be found.[7] Even if this were established, however, antipathy to causal explanation would be likely to continue; for the crucial if not always the explicit objection to talk of causes is that it confuses the domain of the 'physical' and the 'natural' with that of the 'human' and the 'social'. This cherished dualism is threatened by references to the causes of human actions, which references are attacked accordingly as scientistic prejudices, illegitimate extensions of the discourse of the natural sciences beyond their proper realm of application.[8] For our part, we fail to see how sociologists can continue to take this view when they have themselves developed an understanding of science as a form of culture like any other, and decisively eroded the idea that scientific knowledge and scientific discourse are in any important sense distinctive and in need of setting apart. We see nothing at all objectionable in seeking to understand actions by reference to causes, and this is precisely what we shall do.

Causal accounts of the application and extension of scientific knowledge must offer solutions to two kinds of problem. There is the problem of the next case: what was the cause of the exemplar or the classification being applied in that way rather than another on that specific occasion? And there is the more general problem of continuing cognitive order: what causes have acted persistently and systematically over time, so that the repertoire of exemplars and classifications has grown and changed as it did, and not in some other way? Sociological studies will tend to be particularly concerned with the second problem, but even so they need to concern themselves with the first en route, as it were.

What kinds of causes, then, might we sensibly refer to as we seek to solve these problems? First of all, they may not include the alleged intrinsic powers of words and verbal formulations. This is, of course, a basic tenet of the finitist account, but it is worth reiterating, since we can so easily mistake the effects of things mediated by words for the effects of words themselves. If finitism is accepted, then verbal formulations of values, rules, aphorisms and ideas are all alike debarred as causally potent, as are the inherent persuasiveness of rhetoric, or the alleged powers of ideologies or legitimations. Secondly, we need take no interest in causes which operate haphazardly or capriciously, since, although they are no doubt operative, not to say ubiquitous, they will rarely bear upon the problem of continuing cognitive order which is of especial interest in the context of sociology. A scientist identifies substance X as nitroglycerine. Among possible causes of the act are: a brain tumour, an idiosyncratic reading of the instruction book, a personal emotional involvement so overwhelming that it temporarily transforms the individual's intuitions of analogy. And such causes may be operative whether the belief that X is nitroglycerine turns out to be 'true' or 'false'. But circumstances where such causes are sociologically interesting causes will be very rare. Interesting causes are normally those which act persistently over a number of actions, and sociologically interesting ones normally act persistently over the actions of more than one individual. Where we encounter a patterned and recurrent set of identifications agreed by more than one individual, say that substance X is nitroglycerine, it is plausible to expect to find sociologically interesting causation.

What are likely to count amongst the sociologically interesting causes of scientific action are the goals and interests which are furthered by the action being performed in the way it is, rather than in some other way. Similarities and differences may be weighed against each other in innumerable ways, but in particular situations where particular goals are being pursued only one way may be seen as having any point or purpose. Models and exemplars may be modified and adapted in many ways, but in a particular situation just one or a few ways may appear to have any point – just those ways will have any pragmatic validity and count as successful accomplishments (see section 4.3 above). And judgements of this kind are likely to be both persistent and shared.

The suggestion is then that goals and interests are associated with scientific research in all actual situations, and operate as contributory causes of the actions or series of actions which constitute the research. The causes help to solve the problem of the next case, of why a term is applied or an exemplar extended in that particular way that time. Thus it is not knowledge which is treated as caused, but the *change in knowledge* brought about by an action (Figure 5.1). Goals and interests help to explain the change as the consequence of a goal-oriented or interested action. They are not sufficient causes of the action – no verbal account of causation can ever be considered sufficient in any actual empirical situation – but they are causes of the action, which action would be inexplicable without reference to them. Moreover, every action is to be regarded as a caused action: it is not conceivable that an action could have some alternative 'uncaused' form, the form which 'rationality itself' would give it if scientists were 'disinterested', and research in no way goal-oriented. The aim is precisely to account for the

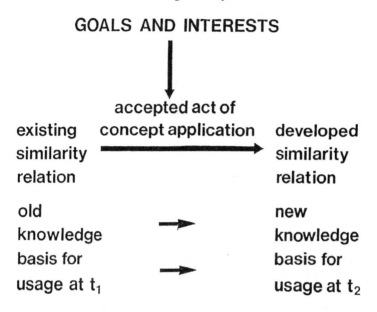

Figure 5.1

rational action of scientists, action which 'of itself' is underdetermined: the claim is that such action is to be understood as purposive and goal-oriented, and that goal-orientation causes the action to be of one form rather than another (Barnes, 1982a, pp. 101 ff).

Aniline reds

For an example we turn to a recent study by Henk van den Belt (1989). He has looked at the development of a similarity relation in a context ideally suited to understanding the process as contingent action – that of a court of law. The court was required to determine which of a number of substances were the same as each other and which different, in order to decide the scope of a patent. The relevant patent law protected substances: it offered their inventors redress against others attempting to manufacture and market 'the same substance'. The substances in question here were the red dyes derived from aniline and first produced, mainly in France, around 1860. A patent granted to Renard Frères in 1859 referred to 'fuchsine' and a process for its manufacture. The court had to decide whether subsequent aniline reds, produced by other manufacturers, breached the patent.[9]

The red dyes of aniline were traded under different names: 'fuchsine', 'azaleine', etc., which in themselves were interestingly patterned similarity relations. The substances referred to by these different names were, however, explicitly recognized as different, and no attempt was ever made to claim that just as they stood they were identical. They were indeed different for important practical purposes; some industrial processes were more effective using the one kind of dye, some using the other. It was a matter of common knowledge that the different processes used to make the dyes produced recognizably different materials.

There was, however, another body of knowledge relevant to the case: that of academic chemistry. Even chemists agreed that the red dyes of aniline were different. But the success of chemistry, both as a version of reality and as a body of instrumental lore, had been based on its readiness to distinguish between essential and incidental similarities and differences, and just this distinction was brought into play here. The research of the chemist A.W. Hoffmann was cited to prove that the colourants present in the various dyestuffs were all salts of the organic base rosaniline. The dyes were thus *essentially identical* in being salts of rosaniline. Indeed there was no need to talk of 'dyes' or 'reds': there was 'the' rosaniline dye, aniline red – a single form of chemical compound, admixed, in the various proprietary substances, with various kinds and levels of 'impurities'. This, of course, was a contestable description; given the importance of positivistic positions in French chemistry it might have been challenged even in a rarefied academic context, but it was certainly challenged here.

Thus the court of law found itself provided with two versions of the relationships between the red dyes. They had, let us say for simplicity, to decide whether fuchsine and azaleine were the same substance or not. And to decide was to decide between opposed versions of nature clearly associated with opposed goals and interests. Renard Frères, its legal representatives, its experts, its supporters, all sought to establish the sameness of the substances. Its competitors, their lawyers, their experts, and their supporters sought to establish difference. The court found for Renard Frères. They balanced similarities against differences, and found the similarities weightier – except that, as these are weightless quantities, this could not have been what occurred. What then do we say of the decision? It was, says van den Belt, "a consequence of the balance of power" (1989, p. 196). This makes the decision 'interested', in the same way as the arguments of the lawyers, experts and so forth were interested; all the assertions of sameness and difference in the courtroom context were linked to goals and interests.[10]

The court may have reached an interested decision, but it was in no clear sense an inferior one – as though the other decision would have been better. Interests conditioned a necessary choice between two reasonable versions of 'sameness of substance'. Was this not, though, merely a legal anomaly, caused by the legislators' forgetting to specify what 'sameness of substances' meant? Not so. The case exemplifies a completely general point, as so many of our previous arguments have been designed to show. Whatever the legislators had specified, they would have forgotten to specify what their specifications meant. Had they specified sameness of colour as the key to the scope of dyestuff patents, lawyers would have debated the same-colour relation. It is a standard point in epistemology that in all description, which consists in one thing being asserted to be the same as others, it is necessary at some point to abandon the attempt to say in what respect they are the same.

What now might be said of the goals and interests operative in our example? First of all, they are social goals and interests, associated not with given individuals but with their place in a pattern of social relationships. The lawyers beautifully symbolize this point: each individual lawyer presented a received brief, but would as happily have presented the opposed brief if circumstances had linked them to it. Secondly, whilst we have two factions with opposed interests, the interests are *not* those of social groups. Individuals of the same group are divided by conflicting interests here, and individuals of different groups are united by shared interests. In each faction a cluster of individuals from different groups, and all with different goals, have an interest in the same judgement because it furthers all their separate goals alike, and is the route to achieving all of them. Renard Frères, their lawyers, their advisors, their supporters all have an interest in the identity of fuchsine and azaleine, since the establishing of that identity will help them all achieve their diverse goals. Thirdly and finally, the interests of the factions as described so far are narrow vested interests peculiar to the factions, and have no general legitimacy as bases for judgements of sameness/difference in nature.

Renard Frères have an interest in establishing the identity of fuchsine and azaleine. It is an extrinsic interest, profit: and one peculiar to Renard Frères – its own profit. Accordingly, if we hear a Frère, or a Frères lawyer or expert, assert this identity, we apply Rice-Davies' law and conclude: well, he would, wouldn't he?[11] Partiality lowers credibility.

But this is, of course, why the testimony of independent experts is of interest to learned judges. Such experts are considered as relatively impartial and disinterested.

Imagine, then, that an independent academic chemist gives evidence on the aniline reds.[12] Such an expert would know that the essential identities of chemistry have justified themselves in terms of general understanding of chemical processes, in inspiring extensions of technique, in increasing prediction and control in the context of the chemical laboratory. He or she would routinely describe the world in terms of atoms and molecules, elements and compounds; as a matter of training, out of habit, perhaps out of expediency in that any other mode of description would be sanctioned by fellow chemists, possibly out of deep conviction that this is the one true account of what the world is like. The dyes would be described to the court on this basis. Such a description would not be the product of judgements conditioned by the narrow extrinsic vested interests of any identifiable faction. Rice-Davies' law would not apply. Nonetheless, the courtroom situation serves beautifully to highlight the fact that this description too is explicable as the product of interested and goal-oriented activities and judgements. For another independent expert, called by the other side, could present the 'technological' account, or the 'positivist chemical' account instead of the 'realist chemical' account. Independent experts, within different bodies of disciplinary practice, with different projects, related to different goals and interests, seeing sameness and difference in different terms accordingly, would offer the court different versions of the truth. Nor is there any stronger sense in which an expert can be independent: in the absence of any project at all, in a situation where goals and interests have no role, the question of the sameness or difference of the aniline reds is undecidable and would be experienced as a meaningless one.

Routine

In the case of the aniline reds the correct extension of a similarity relation was contested, and proved to be highly problematic. Indeed it was more problematic than we have so far indicated, for 'the verdict' was actually a series of verdicts, handed out over a period of years; and the relevant goals, interests and accepted background knowledge all changed over that time.[13] But such problematic extensions of relation of sameness are not typical: we do not normally make choices between alternatives in this way, or even have any awareness of alternatives at all. Much of the time we act automatically and extend knowledge routinely, as a matter of course.[14] We have already recognized that what is routine and matter of course furnishes an essential, ineliminable backdrop against which more innovative extensions of knowledge can take place. But might not the routine and matter of course also be a way of extending knowledge independently of goals and interests, a way sustained simply by habit and authority? Is not the correct use of classifications and exemplars not simply their routine use, their habitual mode of use, and will this not be how usage develops when no goals and interests operate?

Habit and authority do indeed sustain routine usage. Training in routines habituates us to them. We look at the traffic light, and, whilst it tells us nothing, it hits us in the eyeball in a way that elicits the habitual description 'red'. And what we habitually describe as red we know will be so described by others, so that references to red will be collectively underpinned by authority as well as individually by habit. Habit and authority are factors which help us to understand why the light is described as red [along with the nature of the radiation it produces, of the glass it passes through, of the eyes which receive it, and endless other things such as interest physicists, physiologists, psychologists and others]. But although habit and authority help to account for routine action, they do not offer an alternative basis of explanation to that provided by reference to goals and interests, as though action could persist sustained by habit and authority as the only sociologically interesting causes. Rather, habit and authority need to be understood as immediate causes of routine action, which action would be modified if goals and interests required it: in that it persists unmodified, routine action is itself explained by its relationship to goals and interests.

Since scientific controversies have been studied in such detail by sociologists, let us refer to them to illustrate this point. In long-running controversies, inference, problem-solving procedure, classification and even perception become routinized and matter of course on both sides. This makes half a controversy an admirable exemplar of routine knowledgeable activity, but one where the role of goals and interests is also clearly apparent. Everyday classifications and everday knowledge offer similar exemplary illustrations. We identify children as a matter of course in everyday life. We see them around us and refer to them without thought or hesitation. Similarly, and equally as a matter of course, mothers become aware of the unborn children in their wombs; increasingly, they are able to watch their movements on the scanner at the hospital. But here routine usage is variable. In some contexts it is routine to refer to fetuses, not to unborn children. Just as the first usage extends the analogy with paradigm cases of children, so the other limits it, with a usage which may be just as matter of course and unreflective. The two patterns of routine are associated with different clusters of interests, as activists promoting one or other such interest occasionally make dramatically obvious. Given that both patterns of routine are sustained by active reflective human beings, it is reasonable to conjecture that both need to be accounted for by reference to goals and interests, however automatic and 'natural' it may be to act in conformity with either.

Where the goals and interests relevant to a body of routine practice change, the routine practice itself is liable to change. However monolithic the system, however sweetly it operates, however extensive the agreements on which it rests, everything remains revisable. A distinction always remains between what is automatic and routine and what is correct, so that the correctness of routine may always be challenged and hence changed. Indeed such a distinction, which allows automatic responses to be overridden, is actually essential not just to allow systematic change in a system but, as we argued in Chapter 4, to sustain the system as well. Thus, explanation of the development of scientific practice by reference to goals and interests does not deny that concepts and exemplars may be applied routinely as a matter of course: indeed it requires it; it makes it a part of the account of how goals and interests operate. But in treating it in this way it does indicate that practices cannot be accounted for in their persistence simply by noting that they are routine, or entrenched. Within any community the existence of some degree of shared routine is essential: it constitutes automatic agreement in practice essential for communication and exchange of information. But any part of agreed practice may be made the object of reflective evaluation (wherein much of the rest of agreed practice must be temporarily taken for granted). And the outcome of the evaluation is in need of explanation. If a routine is changed we need to consider why it is changed. If it continues to be accepted we need to consider why it is not changed – why what is routine is allowed to count as what is correct.

Individuals in all forms of life are able to extend knowledge automatically, without hesitation, as a matter of course. This is how most initial designations of things are made. And although such an initial designation is never sacrosanct, every individual who makes such a matter-of-course designation will assume that it is also what would come naturally and automatically to everyone else. The initial response is the 'prominent solution' to the problem of linguistic coordination (Schelling, 1960; Lewis, 1969).

In all contexts where agreement itself is more important than the specific form that agreement takes, this solution will stand. A principle of maximum cognitive laziness will appertain: other things being equal, allow what is routine to count as what is correct. But if other things are not equal, if the extra work is pragmatically justified, if a specific kind of agreement is particularly valuable, then automatic tendencies may be overridden. New patterns of coordinated linguistic practice may be established, just as they have to be when there is no existing agreement in routine practice in the first place.

The machinery involved in the perception and recognition of things hums along undisturbed much of the time. For the individuals in a given culture it usually hums along in unison; indeed it has so to do for the culture to exist. The fact of its existence depends upon a certain blind conformity in perception, understanding and judgement, in *initial* responses to things. But the machinery of perception and recognition is nonetheless subordinate to reflection and calculation. The basis of sociability, and thus of humanity, lies in our shared tendencies to automatism, but its actual achievement lies in the calculative exploitation of these tendencies.

5.3 THE EXPLANATION OF SCIENTIFIC CHANGE

Significant long-term changes in bodies of scientific knowledge are complex historical processes made up of innumerable specific actions, each involving the problem of the next case. For each particular action, existing knowledge stands as a resource, a set of accepted precedents awaiting further use/extension as the result of a contingent judgement. The knowledge itself is inert, simply a cluster of existing instances. It is an agent who links the cluster to the next case, using the knowledge, adding to it, changing it, in a single act. Many such acts constitute the growth of knowledge. We must imagine the problem of the next case being solved over and over again on a grand scale in sequences of unfolding actions. On all the actions, goals and interests will have their effect, sometimes in one arrangement, sometimes in another, sometimes with one kind of outcome, sometimes with another. The full details of processes of this kind must remain beyond the grasp of even the most conscientious historian, but it is reasonable to look for general trends and tendencies. Sometimes the effects of goals and interests will interfere or cancel out; sometimes they will reinforce each other. Those goals and interests most consistently and persistently involved as scientists extend and modify their repertoire of exemplars are likely to be the most important in understanding any general patterns in what has occurred.

One of the interesting findings when scientific change is studied in this kind of way is how often goals and interests of an illegitimate, credibility-reducing kind are operative. When scientists extend their knowledge to bear upon issues involving practical decision making it is well known that they are liable to tailor it to support the policy outcome they wish to advance – indeed those who study these issues suggest that this is the typical fashion in which science enters into the arena of practical policy (Nelkin, 1975, 1979; Collingridge and Reeve, 1986). But even in the esoteric context of 'pure' discipline-based scientific research the modulation of judgements and practices by exoteric political objectives and interests is well documented; and the analogous role of professional vested interests, whether of disciplines or specialties or laboratories or even national scientific communities, is even more clearly established. Moreover, the causal involvement of interests and objectives of this kind in processes which have produced currently accepted, indeed currently venerated knowledge, is evident – as we should expect, since the causation of these processes, even by 'inappropriate' interests need have no bearing whatsoever on how their outcome is subsequently evaluated (Shapin, 1982).

The considerable body of work on what may be described, for brevity, as the role of unacknowledgeable political and ideological interests in scientific research is in many ways of great importance. It enriches our understanding of the historical development of science and (yet again) calls into question the unduly idealized pictures provided elsewhere. However, there is an unfortunate tendency for references to *illicit* interests to be perceived as what a sociological orientation to science fundamentally involves: the assumption that sociological explanation is applicable only to what is erroneous, irrational or illegitimate remains stubbornly resistant to eradication. And this means that studies of narrowly conceived 'technical' activities, linked to legitimate and acknowledged interests in prediction and control, serve better to illustrate the fundamental characteristics of sociological explanation than any of the equally fine studies of the role of exoteric political interests in scientific research.

The fundamental point is this: identify an episode of scientific change by whatever criterion or method you will, and it will be a series of contingent actions modulated by goals and interests. Mary Hesse offers an idealized picture of science as a process of inductive learning shaped and evaluated in terms of a general 'pragmatic criterion' of increasing prediction and control. The standing of such an idealization is obviously problematic: who is to say why this rather than some alternative should be taken as our criterion of good science? But suppose we set this aside, and try to identify some particular scientific change by use of this picture, some episode in the course of which scientists got better at predicting certain things or manipulating things to a desired outcome.

The point is *not* that we could not hope to find any such science. We could. Scientific activity does proceed by inductive learning, and can be (successfully) directed to achieving improvements in prediction and control. The point is that the activity would *at the same time* be found to be conventional activity modulated by goals and interests. Inductive learning always goes one way rather than another equally 'logically possible' way, and the scientific actions we have been discussing have been the actions which take it one way rather than another. When people learn they do not learn 'by induction' pure and simple: they rather constitute an inductive learning machine tuned to convention. The outcome of its operation is never increased prediction and control *per se*, but increased utility of particular cultural resources for particular kinds of prediction and control as described in a particular form of shared discourse. With these points in mind, let us now turn to an example of large-scale scientific change.

The zymase controversy
In Robert Kohler's investigations of the emergence of biochemistry in the first two decades of this century, we find a detailed account of its establishment as a scientific discipline, of its central theoretical commitments and research programme, and of the controversial career of one of its key exemplars, zymase and the demonstration of its ability to induce cell-free fermentation (1972, 1973).[15]

The emergence of biochemistry involved, not a simple combination of the existing methods and theories of two disciplines, but the acceptance of a new theoretical orientation and a new research programme. From the 1860s through to the end of the 1880s the dominant account of cellular operation was the protoplasm theory, which attributed chemical reactions within the cell to the powers of a single homogenous substance, living protoplasm. The emergence of biochemistry in the

1890s coincided with the rise of the enzyme theory, which related every particular chemical reaction to the operation of a specific enzyme.

To attribute the chemistry of the cell to undifferentiated protoplasm was to accept that the cell as a whole was necessary for that chemistry to occur. This position was widely felt, prior to the 1890s, to have been established decisively by the experimental work of Pasteur. He had shown that to kill living cells was to terminate the characteristic chemical processes associated with them; and, like others, he had failed in careful attempts to extract chemically active fluids from cells. Although this work never went entirely undisputed, its dominance was reinforced by the failure of Pasteur's critics to accumulate positive evidence for a 'purely chemical' theory of cellular processes. It also gained credibility as the basis for a number of successful techniques based on the equation: specific reaction = specific micro-organism. These techniques allowed the improvement of industrial processes through the careful identification and selection of distinct strains of micro-organisms. Finally, the protoplasm theory could be made to fit nicely with a range of vitalist and holistic positions in the biological sciences, and with the anti-reductionist and anti-materialist sentiments of important constituencies both within and beyond that context.

The protoplasm theory was nonetheless abandoned with remarkable rapidity, so that from being orthodoxy in 1890 it was struggling for legitimacy in 1900, and in 1901 Franz Hofmeister could gain widespread recognition for his claim that 'sooner or later a particular specific ferment will be discovered for every vital reaction' (Kohler, 1973, p. 185). Kohler presents Hofmeister's lecture on 'The Chemical Organisation of the Cell' as a manifesto for the new discipline of biochemistry and an assertion of its central dogma, that not a specific micro-organism but a specific enzyme (ferment) should correspond to every particular chemical reaction in biological systems.[16]

The enzyme theory was by no means a new development at the end of the nineteenth century. It had existed for several decades, and had been accepted as compatible with and supplementary to the protoplasm theory. Enzymes were known to be secreted by cells and to catalyse simple chemical processes like hydrolysis or the decomposition reactions involved in digestion. But no evidence had been found that the complex chemical processes of the cell itself depended upon the presence of specific enzymes. During the 1890s, however, the situation was transformed. In 1894, Fischer established the stereochemical sensitivity of enzymes, their ability to affect one kind of molecule but not its stereochemically equivalent isomer. This was testimony to the high level of discrimination and specificity of action of enzyme catalysts, and was bound to add to the credibility of the one enzyme/one reaction doctrine. In 1898, Croft Hill offered experimental evidence that enzymes could catalyse chemical synthesis as well as decomposition, just as the theoretical accounts of catalysis being developed by physical chemists suggested. This permitted the enzyme theory to be extended to chemical reactions of any kind and the production of molecules of any degree of complexity. In 1897, Edward Buchner showed that extracts from pressed yeast cells would decompose sugars into alcohol and carbon dioxide, a capacity hitherto held to be that of the intact living yeast cell itself, and never before observed in the absence of such cells. This immediately made a reductionist programme, involving a renewed search for enzymes in the internal fluid environment of the cell, more plausible.[17]

Kohler's account of the subsequent triumph of the enzyme theory strikingly exemplifies many important sociological themes, not least our earlier suggestion that it is impossible in practice to identify what precisely a theory consists in. Kohler emphasizes that the idea of an 'enzyme' or 'ferment' was vague, and that in so far as it had a meaning that meaning was varied. Traditionally, an enzyme was an active agent found in *extra*-cellular fluids like digestive juices. To refer to an *intra*-cellular enzyme was initially close to being a contradiction in terms. But this modification of reference was crucial to the increased significance which 'the enzyme theory' came to enjoy, and was accomplished by the scientists involved without difficulty. Kohler also calls attention to the fuzzy boundary between the enzyme theory and the protoplasm theory. Initially, it was possible to see the enzyme theory not as an alternative to the protoplasm theory but as a development within it. Enzymes could be regarded as the products of protoplasm, the means by which protoplasm accomplished certain kinds of chemical change, the mediating factors linking it to highly specific chemical reactions. At a yet more speculative level, enzymes could be understood as detached sidechains of large and complex protoplasm molecules, and hence actually as pieces of protoplasm itself. But, as time went by, what were regarded as the promising implications of theory in this area were increasingly considered implications of 'the enzyme theory', and 'the protoplasm theory' was equated solely with what previously had been its least applicable and most suspect aspects – especially those implied by the term 'living protoplasm'. Although

Kohler describes a transition from 'the protoplasm theory' to 'the enzyme theory', his narrative fits equally well with the notion that 'protoplasm theory' split into two or more differentiated variants, and that one of these was increasingly referred to as 'enzyme theory'.

Kohler's own opinion of the enzyme theory is that 'properly speaking [it] was hardly a theory at all' (1973, p. 186). Reference to 'enzyme theory' is only justifiable, he tells us, because the more accurate term 'enzyme program' is 'so clumsy' (1973, p. 186). The rise of 'the enzyme theory' was actually the rise of a research programme directed towards 'the isolation, purification, and physico-chemical study of enzymes for all the known biochemical functions' (1973, p. 186). Commitment to the programme was commitment to the study of cellular processes as chemical reactions in an aqueous medium, analogous to the familiar test-tube reactions already addressed by standard procedures in existing branches of chemistry. The promise and feasibility of this programme had become recognized through the development of new exemplary techniques and procedures: participation in the programme was involvement in the extension and development of the techniques. In effect, therefore, the 'rise of the enzyme theory' was actually a reorientation of practice around a newly available set of exemplars.

The exemplar chosen for extended discussion by Kohler is provided by zymase, the first enzyme shown to catalyse a complex chemical reaction without the environment afforded by the living cell. The key to the 'discovery' of zymase lay in the invention of a method for the extraction of significant quantities of intra-cellular juices. This problem, which had defeated many earlier investigators, was solved by Martin Hahn in 1896. It was well known at this time that prolonged grinding with sand would break the walls of many kinds of cell and permit an outflow of their juices, but the separation of juices from the debris of the cells had proved a stumbling block. Hahn solved the problem for ground yeast cells by mixing them with kieselguhr, and crushing the resulting material in an hydraulic press. The technique was developed to further research on intra-cellular proteins, and it led to the identification of zymase only accidentally. Hahn had found his juice extracts to be unstable, and investigated the effects of various known preservatives upon them. One reagent he tried was 40 per cent glucose, which was added to yeast juices shortly before Hahn took his vacation. During this period the 'discoverer' of zymase, Edward Buchner, visited the laboratory and noticed gas bubbling from the 'preserved' juices. He recognized the process of fermentation of glucose into alcohol and carbon dioxide, and, having confirmed to his own satisfaction that residues of living yeast cells were not responsible, proceeded to announce the first clear instance of non-hydrolytic cell-free fermentation (Kohler, 1971).

Buchner's Nobel-prizewinning discovery was indeed a novel observation and one of great theoretical importance: it long served as the standard demonstration that intact protoplasm was not required even for the most complex metabolic processes, and as a key event in the mythology which celebrated the emergence of a mechanistic and reductionist approach to biology. But even more important, it stood as an exemplary achievement: the use of the crucial extraction technique and the classification of the resulting extracts in terms of biochemical function led to the identification of a considerable number of endo-enzymes, as well as other advances in scientific understanding of cellular processes. Further research could be modelled on the zymase work, and it was.

Nonetheless, it would be wrong to claim that the procedures of Hahn and Buchner were adopted simply 'because they worked'. A lengthy controversy characterized the reception of zymase. Whether the described techniques worked was called into question, and what followed given that they did work was subject to different interpretations. Many scientists recorded their inability to replicate Buchner's results. Others linked them to a failure to eliminate fragments of protoplasm from the yeast extract; and, when Buchner demonstrated that treatment with cell-killing antiseptics failed to destroy the activity of the extracts, they questioned whether that which killed a living cell need necessarily destroy the functioning of cell protoplasm. Others again refused to admit any radical implications, even if the results proved replicable and the experimental procedure which produced them was fully competent. The observed extra-cellular reaction might always turn out not to be the same as the familiar reaction associated with the intact yeast cell, which latter reaction might require the mediation of protoplasm after all. Or again, the extra-cellular reaction might be the same as that within the cell in a sense stronger than Buchner was willing to admit, in that it could be a reaction caused by invisible or even soluble elements of protoplasm itself: what Buchner called an enzyme could actually be more like protoplasm than it was like standard instances of enzymes, and indeed Buchner himself had inadvertently offered support for this interpretation by acknowledging difficulties in describing zymase entirely by analogy with these existing instances.

A familiar problem is recognizable here. In order to assimilate Buchner's work, scientists had to extend and modify their existing sense of what was the same as what in nature. But there were many ways of extending that existing sense, corresponding to many equally reasonable ways of weighing similarity against difference. Those ways actually encountered can be understood only as contingent historical actions, and explained, in so far as it is possible to explain them, only as the consequence of pragmatic considerations.

Kohler himself attempts to understand and explain some of the responses to Buchner's work in this kind of way. He links different responses to different established groups of scientists and their particular pragmatic concerns. One intriguing feature of these responses is how strongly future-oriented they show the scientists to have been, and how little constrained and restricted either by their existing theoretical inclinations or, perhaps more surprisingly, by their existing repertoire of skills and competences. The idea that commitment to tradition or even commitment to 'form of life' can be made the basis for the explanation of scientific judgement in this controversy is not at all easy to defend. The notion that categories and classifications are hardest to change amongst those in whose practice they are most deeply entrenched faces difficulties.

Consider the scientists identified by Kohler (1972) as constituting the most formidable initial opposition to zymase. The 'technologists' were scientists closely involved in the invention and refinement of industrial and agricultural processes, successfully extending accepted techniques for the selection and cultivation of cells and unicellular micro-organisms: many of them were involved with brewing, oenology or mycology. Their specific skills and competences required no observation, manipulation or even speculation at the sub-cellular level, and no knowledge of the internal chemistry of the cell. Their practice was described, reflected upon and legitimated in the vocabulary of cells and protoplasm. Their tradition was recognized as deriving from Pasteur.

As far as zymase is concerned, there is no difficulty in relating the initial strongly negative response of these scientists to their existing habits of thought and 'theoretical commitments'. Nor is it implausible to see the cell-busting approach of the enzyme programme as a threat to their own established practice. Nonetheless the group is an especially interesting one, for whilst it provided the main original opposition to Buchner's claims, its members soon came to figure amongst his strongest supporters. The grounds they themselves offered for their rapid change of opinion were straightforward. At first they had failed to replicate the zymase experiment, but later, with practice they had succeeded: that was enough for them.

It cannot be enough, of course, for anyone seeking to understand and explain what the technologists did. They did not have to take their own empirical work as replicating Buchner's. They did not have to describe their empirical results as confirming Buchner's theory. Other scientists acted differently in both respects, and gave good grounds for doing so. Perhaps the technologists were predominantly practical and instrumental in their orientation, and saw the chance of making use of Buchner's techniques. Perhaps they were attracted by the possibility of enzymatic brewing, and thrilled to the prospect of cell-free reagents implicit in zymase. But Kohler is careful to point out that, if this was indeed the case, then it was entirely promise and not at all pay-off which was involved. Only after the First World War did solutions of enzymes find any practical application; the technologists had to reconcile their enduring belief in zymase with continuing reliance on their long-established techniques and procedures.

The technologists were not the only supporters of zymase closely associated with the tradition of research it apparently threatened. The most immediate and enthusiastic support for zymase came from Paris, from some of the disciples of Louis Pasteur and senior figures at the Pasteur Institute. Kohler shows that this is not in itself surprising, that these were people whose concepts and techniques were already close to those of Buchner, and whose research in immunology, immunochemistry and enzymology already involved their acting and thinking at the sub-cellular level. But how was it that this was so? How had this kind of science emerged out of a Pasteurian research programme which had initially used the same methods of selection and manipulation of intact cells as those the technologists employed in their work on industrial processes? There is a problem here only if a research tradition is mistakenly assumed to restrict, narrow or constrain scientific activity. Once it is accepted that the resources of a tradition serve to facilitate further activity, the vast repertoire of know-how in the manipulation of populations of cells can be recognized as an invaluable springboard for further work at the sub-cellular level.

As an aside from the mainstream of Kohler's own argument, which confines itself to the institutional context of professional science, it is worth noting that additional incentives for a move in the direction represented by the enzyme programme, and away from some of the key tenets attributed to a 'Pasteurian' programme, are evident in the exoteric political context in France. The time of Pasteur's greatest success coincided with the zenith of the Second Empire, when his holism and anti-materialism did him no harm as a recipient of patronage from a polity very closely allied with the Catholic Church, in a system where the government relationship with science was mediated by a 'Minister of Education and Religions'. From 1870, however, the political context was that of the Third Republic, and although the ideological possibilities of science were of as much political interest as ever, it was a different ideology – more or less the opposite ideology – for which there was a demand. And where there was a changed demand it is reasonable to expect a changed supply, particularly in light of the fact that 'French scientists have a remarkable record of being faithful servants of the state and have had little difficulty in changing allegiances to regimes' (Paul, 1985, p. 22). Even in the public pronouncements of Louis Pasteur himself, so Farley and Geison (1974) have sought to demonstrate, the emphasis on vitalism and holism declines from the heyday of the Second Empire, and materialist themes and arguments become increasingly evident. But this is to identify possible causes of a kind we wish as far as possible (at present) to leave out of account, so let us return to Kohler's own narrative.

Kohler (1972) employs a familiar explanatory scheme which has often been used in the sociology of knowledge. He identifies an established group of scientists with specific interests and refers to those interests to help to explain the response of the group to zymase. But whilst the doctrine – one group, one set of interests, one response – represents his main explanatory strategy, his historical narrative moves well beyond what can be encompassed within it, and reveals it as at best a pragmatically useful simplification. Thus, Kohler shows that in physiological chemistry, the main discipline involved in the controversy, there developed a great range of responses to Buchner's claims, responses which exploited in all manner of ways its repertoire of cultural resources and accepted knowledge: for a physiological chemist Buchner's work could imply that the protoplasm theory was dead, or that zymase was alive, or any number of intermediate positions. Conversely, Kohler identifies similar responses emanating from individual members of quite different scientific groupings. Indeed the major concern of his account of the zymase controversy is to show how support for zymase came from a number of different institutional locations, from members of a number of existing disciplines and specialties.

Nonetheless group affiliation remains Kohler's major explanatory resource in accounting for variations in scientific judgement, and his response to the complexities of his own narrative is to extend the utility of that resource by thinking in terms of a latent group. Buchner's supporters were alike in showing what would later be recognized as a biochemical orientation. 'Zymase was . . . a touchstone of pre-existing opinion'; 'The zymase debate enabled the emerging group of biochemists to recognise themselves as a group with a community of outlook and objectives' (1972, p. 351). Looked at in this way the interests and outlook of the 'biochemists' might help to account for their scientific evaluations. But there is a danger of running causation backwards here, and using factors evident only after the acceptance of zymase to account for that acceptance: at the time of the controversy all that clearly distinguished the group of future biochemists seems to have been their favourable evaluations of zymase itself. The role of references to biochemists and biochemistry at this time is beautifully characterized by Kohler, borrowing the words of Fritz Ringer: 'terms may become fashionable because they permit their users to regard each other as allies before having fully understood one another' (1973, p. 194).

What we have in Kohler's case study are changes in knowledge and competence, and changes in social relations and groupings, running together as one and the same process. In due course, zymase becomes an exemplary achievement, with a particularly important role for scientists who have come to identify themselves as biochemists, working in biochemical laboratories and publishing in biochemical journals. But when zymase was actually becoming an exemplar biochemistry was becoming a professional grouping. And this is not plausibly describable as an 'already there' biochemistry becoming self-aware. It is rather a 'not already there' biochemistry taking shape, coming into existence, in and as the interactions which were the technical debate. The process of cultural change was also a process of group formation.

In a situation of this kind the simplifying strategy of accounting for whole sets of scientific judgements in terms of the characteristics of fixed groups in definite institutional locations has little or no pragmatic value, and must be discarded. Always a simplification, here it is too much of a simplification. Contrary to what is sometimes assumed, it is no part of the project of sociology to explain every human action as that appropriate to some given group, or status, or institutional location, or to account for an unfolding series of actions wholly externally, with no regard for the emergent connections between them.[18]

What is required here is a stricter application of the finitist position, one which recognizes cultural change as a myriad of particular actions, performed not in a fixed context of given groups with given goals, but in a changing context constituted in part by the unfolding actions themselves. This will not atomize history into a succession of unconnected unique events, but will allow the identification of pattern and order of the course of change itself. Simply to speak of culture, even of changing culture, is to presume the existence of order. Controversy itself is orderly and involves the search for greater order: why else would there be controversy? Indeed, when a controversy like that over zymase becomes discernible, a movement toward order has been made: whether it is one position, or two, or a few, which stabilize from innumerable possibilities, it represents an achievement for social cooperation, a successful coordination of individual actions and judgements which secures a measure of coherence in practice.

In the zymase debate, we have an initial stage in which a great number of accounts are advanced, and many different empirical findings are reported. The very existence of these accounts and reports, and their association with particular scientists or laboratories then becomes one of the most important features of the situation in which subsequent debate occurs. Scientists present new empirical work as compatible with this existing account and awkward for that, mindful both of the technicalities of what they seek to describe and of the political expediency of supporting this view or opposing that. Existing positions are made out as mutually compatible or mutually exclusive by scientists aware again both of the consequences for technical practice and the opportunities of political alliance. By the close of Kohler's narrative, these social processes have resulted in the 'biochemical' account of zymase becoming close to predominant, whereupon it is that very predominance which comes to serve as the predominant basis of its credibility.

At this point the original technical issues recede into the background and the technical advantages of zymase become increasingly advantages of adoption.[19] That which enjoys the greatest level of current acceptance is the best basis for communication and interaction, for coordinated, cooperative research, for learning from others, probably for keeping on the safe side of the major powers in the field. Thus, for many scientists zymase is accepted as a prominent solution to the problem of which exemplary achievements to get behind, which to use and extend, which to enjoin upon others. In solving this problem of coordination, these scientists complete the process which has been establishing 'zymase' as part of the observation language of science, and themselves as a part of its social order. And in the sequence of actions and judgements of which this is the outcome, causal connections are clearly discernible between the earlier and the later.[20]

CHAPTER 6

Drawing Boundaries

If the finitist account of concept application is correct, then it will apply to the concept of 'science' and of 'scientific knowledge' just as it applies to any other concepts or categories. What is to count as either of these things will be a matter of agreement, and it will be revisable. In so far as these concepts are used, these instances of use will be matters of fact to be understood in relation to the contingencies of particular historical situations. Similarly, the boundaries between scientific disciplines and specialities will be contingent accomplishments originating in specific situations and liable to revision as these situations change. The aim of this chapter is to study such boundary drawing by scientists as contingent social activity. The demarcation between what is scientific and what is not will be considered first and then we will look at demarcations between 'sub-disciplines' within science: a case study will be the basis of discussion in each case.

6.1 DRAWING THE BOUNDARIES OF SCIENCE

The scientific profession possesses considerable cognitive authority in modern societies, and indeed wherever 'science' is identified and designated as such the implication is that something especially trustworthy or reliable is being described. Such authority is of course of inestimable value to individual scientists, and they have a vested interest in its maintenance. They can be expected to police the existing boundaries of science, to avoid the intrusion of whatever may detract from its reputation and to seek to expel anything potentially disreputable which arises within it.

Nonetheless, the methods and criteria employed in these endeavours clearly change from one historical context to another, and in consequence exemplars of legitimate science are historically variable. Thus, work on N-rays was considered genuinely scientific in France, while scientists of other nationalities were united in dismissing it as fraud or figment (Nye, 1980). The practice of acupuncture has a 'scientific' rationale in China, but in the West it remains, at best, a curious if efficacious empirical technique. Social Darwinism, once thought to be a routine extension of Darwinian and other scientific evolutionary ideas to mankind, has since been denied the status of real scientific knowledge. Astrology, once indistinguishable from astronomy, and homoeopathy, which for a while after its inception held real promise of becoming the orthodoxy in medicine, remain firmly saddled with the label of pseudo-sciences in spite of recent work which seems to some to call for a reassessment (Gauquelin, 1984; Benveniste, 1988).

Michel Gauquelin's statistical evidence in support of astrology would perhaps be a serious embarrassment to scientists if they were not so good at ignoring it. But one day it could conceivably come to be accommodated as a triumph of the scientific method. Gauquelin's work seems to imply the existence of forces and interactions unrecognized by current scientific theory and yet it is based on methodological principles and empirical evidence which have so far stood up to sceptical challenge. Two other examples, only marginally beyond the pale, are acupuncture and water dowsing; these seemingly empirically effective techniques have been excluded from the ranks of legitimate science because they are incompatible with prevailing scientific theory. They could 'in theory' be transformed into instances of 'good science', either by a modest shift in our conception of what constitutes science in the direction of good method or of efficacious knowledge, or by a theoretical reformulation which reconciled current science with the knowledge in question.

In some ways Gauquelin is fortunate in merely being ignored; many practitioners of 'genuine science' fare no better. Other producers of 'suspect' work which nonetheless has prima facie claims to be scientific encounter active hostility. Parapsychology and creationism, for example, continue to be attacked and stigmatized as pseudo-scientific, and the 'pretensions' of their practitioners are often ridiculed. Nor is the way that such fields are discriminated from 'genuinely scientific' enterprises invariably fair and even-handed. Often historically specific criteria of good science are selectively applied, and 'pseudo-sciences' are condemned as such on the basis of tests which most currently accepted genuine sciences would surely fail (Collins and Pinch, 1982).

It would be an unfortunate misreading of the arguments in this chapter, however, if they were taken merely as criticisms of scientists' methods of maintaining professional and disciplinary boundaries. There is indeed a case for saying that these methods often deserve criticism, that scientists' evaluations of fields such as parapsychology are often impossible to justify *even in terms of their own preferred standards*. But formally speaking this is a small point. The large point, on which the present argument is focused, is that, however carefully and even-handedly scientists apply demarcation criteria, what they are engaged in is contingent social action which must be set in its historical context and viewed from a sociological perspective if it is to be adequately understood. At any time there are various criteria which are generally regarded as legitimate bases for demarcating science; at present, for example, it is acceptable to suggest that genuine scientific knowledge should be consonant with observation and experiment. It is necessary that scientists deploy such criteria, and relate their evaluations of knowledge claims to them. This can be done in ways which would generally be considered scrupulous and reputable. When it is so done the results tend to be accepted as legitimate and perceived to be of great practical value and utility. Nor need historical and sociological elucidation call into question that legitimacy and utility. Nonetheless, the demarcation criteria must be regarded as conventional, and their application in all cases as situated human action.

Clearly, then, a case study of the use of the criterion of 'consonance with experiment and observation' would be particularly suitable to illustrate our key point and develop the fundamental argument. Something very close to this does indeed exist in Steven Shapin's and Simon Schaffer's *Leviathan and the Air-Pump* (1985), a work with a whole range of deeply interesting implications for the sociology of scientific knowledge. In *Leviathan and the Air-Pump*, Shapin and Schaffer focus upon some of the scientific work of Robert Boyle, and upon his advocacy and exemplification of 'the experimental method'. Boyle is often remembered today as a key figure in the elaboration of experimentalism but, as Shapin and Schaffer point out, the experimental method was not initially regarded as crucial to obtaining knowledge of nature. In fact, Boyle's promotion of this methodology gave rise to vigorous controversy, and detailed analysis of the controversy may yield an explanation, though not a justification, of his victory.

Robert Boyle, the seventh son of one of the wealthiest men in England, the Earl of Cork, came to be widely regarded by contemporaries at home and abroad as the leading natural philosopher in England. He was one of the prime movers in the reform of natural philosophy in the Restoration period. A founder member of the Royal Society of London, he used his personal prestige and influence to mould this earliest of scientific societies into the institutional embodiment of his reformist ideas. The intellectual leaders of the Royal Society, inspired in large measure by Boyle, presented it as a forum for establishing incontrovertible truths and for settling controversies. Shapin and Schaffer focus on Boyle's experiments with a specially constructed air-pump, conducted with the help of his one-time assistant and then Curator of Experiments at the Royal Society, Robert Hooke. This work in pneumatics is notable not only because it led to the discovery of Boyle's Law but also because it generated an intense controversy with one of the leading philosophers of seventeenth-century England, Thomas Hobbes.

Hobbes's reputation today rests almost entirely upon his work in political theory, but in Restoration England he was equally well known as a mechanical philosopher and author of a system of natural philosophy to rival that of the great French mechanical philosopher, René Descartes. Nobody at the time was in any doubt, however, that Hobbes's political theory and his mechanical philosophy were all of a piece. Boyle himself explicitly stated that his refutation of Hobbes's physical theories was intended to undermine his religious and political views: 'it might possibly prove some service to higher truths than those in controversy between him and me, to shew, that in the Physics themselves, his opinions, and even his ratiocinations, have no . . . advantage' (Boyle, 1772, i, 187). Shapin and Schaffer's examination of the Hobbes–Boyle controversy, as Latour has pointed out, reveals the political theory implicit in Boyle's science, just as it reminds us of the importance of Hobbes's scientific ideas to his political theory (Latour, 1993, pp. 16–17).

Boyle sought to reform natural philosophy by developing a new method based on experimentalism. Before this time the experimental method was associated traditionally with alchemists, and more recently with various 'new' philosophers who used specially designed experimental demonstrations to prove the truth of their preconceived hypotheses (a method that today might fall under the rubric of the hypothetico-deductive method). Francis Bacon had already, some fifty years before, dismissed these narrower motives for experimentation as biased and liable to lead inevitably to confirmation of whatever hypothesis was under consideration.[1] He suggested a new kind of empiricism which sought only to generate new facts. Bacon advocated the gathering of, supposedly, theory-free factual data in 'natural histories' compiled by the collaborative efforts of trustworthy individuals. The Fellows of the Royal Society presented themselves as Baconian reformers, engaged on just such a collaborative enterprise, and Robert Boyle signalled his own commitment to Baconianism in the titles and the form of his many published works: *The Natural History of Colour*, the *Natural History of Heat and Cold*, and so on. This new Baconian kind of experimentalism, ostensibly eschewing all theorizing and speculation, enabled Boyle and his colleagues in the Royal Society to insist that the knowledge which they presented in their accounts of the physical world was concerned solely with demonstrable matters of fact, and was free from any bias introduced by preconceived hypotheses.

The significance of this, for Boyle and his colleagues, was that facts could not be disputed and provided, accordingly, the foundation for universal assent. To go beyond discourse about facts was to invite controversy since hypotheses were always open to dispute. One of Boyle's favourite metaphors likened nature to a vast and complex clock. Experiment and observation show us without any ambiguity the effects of the clockwork – the position of the hands, say, or the pattern of its chiming – but the clockwork itself, the precise means by which the hands are turned or the chimes made to sound, might be various. The metaphor nicely sums up Boyle's concern, in all his scientific work, to demarcate what can be known from what can only be hypothesized. Such a demarcation in turn provided Boyle with a means of differentiating disputes which could be resolved from those which could not. Disputes which could not be resolved, because they were concerned with the validity of a particular hypothesis, were not scientific disputes at all, according to Boyle, but were designated as 'metaphysical'. This enabled Boyle to dismiss at a stroke – and thus to resolve – a number of ongoing debates. For example, although the overwhelming majority of Boyle's contemporaries believed in the particulate nature of matter, and that all bodies were made up of specific combinations of invisibly small particles, there was no agreement as to whether the particles were indivisible, as the atomists believed, or infinitely divisible, as Aristotelians and the followers of the French philosopher, Descartes, believed. For Boyle, this was a metaphysical dispute, incapable of being decided by experimentally generated facts.

Boyle's attempts to create clear demarcation criteria for science were reinforced by other members of the Royal Society. According to Thomas Sprat, author of what is effectively the Society's manifesto, *The History of the Royal Society* (1667), the Society as a whole rejected metaphysical issues and eschewed theorizing in favour of establishing matters of fact. Indeed, he explicitly claimed that a major function of the Society was to decide upon matters of fact. The purpose of the 'Assembly', Sprat wrote, having collectively witnessed an experiment was 'to judg[e], and resolve upon the matter of Fact' (p. 99). It is surely an indication of the importance of the boundary work which was going on at early meetings of the Society that Sprat described the meetings in these terms. The sense conveyed, revealingly but perhaps inadvertently, by Sprat is not that the assembly simply witnesses an obvious matter of fact, but that those present witness an experiment and the precise matter of fact revealed in this experiment is then negotiated and collectively decided upon.

Boyle's and the Society's claim that experiments produce matters of fact depended for its validity upon the assent of others in the relevant community. When Sprat wrote of the 'Assembly' judging and resolving 'upon the matter of Fact', he was trying to convince his audience of the reliability of factual reports from the Royal Society. 'In matters of life, and estate', Sprat pointed out, people 'expect no more, than the consent of two, or three witnesses', with regard to the knowledge produced by performing an experiment before the Royal Society, interested parties 'have the concurring testimonies of *threescore or an hundred*' (p. 100). Similarly, Boyle wrote up his own reports of experiments in such a way as to make possible what Shapin and Schaffer refer to as 'virtual witnessing': the production in a reader's mind of so convincing an image of the experimental performance that direct witnessing or replication is regarded as entirely unnecessary (Shapin and Schaffer, 1985, pp. 60–5). To this end Boyle developed a new and extremely influential style of scientific reporting: prolix and almost overburdened with circumstantial details. The collective witnessing of an experimentally generated fact, then, was regarded as a way of gaining the assent of the community that such a fact had indeed been produced. It was the consensus which guaranteed the factual status of the knowledge in question. Without such a consensus the fact could not be said to have been established.

The credibility of a collectively witnessed fact did not depend solely upon *numbers* of witnesses, however. Boyle and the leading figures in the formation of the Royal Society were concerned to establish and maintain a universally recognized relevant community of scientific observers and experimenters. The relevant community was composed of witnesses whose word, even when taken singly, was generally recognized (at the time) to be trustworthy. The Royal Society, in keeping with its Baconian precepts of collaboration, invited reports of factual matters from all quarters, including merchants, sea captains and even artisans. Moreover, the Society defended the reliability of these sources of information on the grounds that such witnesses would be unlikely to colour their observations, or their reports of observations, in terms of Aristotelianism, as might be the case with any university-trained observer, for example, or in Cartesian terms, as might be the case with anyone well read in current natural philosophy. But use of such sources, and apologies for them, was comparatively rare. For the most part the Royal Society chose to emphasize the fact that of its members, 'the farr greater Number are Gentleman, free, and unconfin'd' (Sprat, 1667, p. 67. See also Shapin, 1988, p. 396; Dear, 1985, pp. 156–7).

It was the gentlemanly status of the majority of the witnesses in the assembly of the Royal Society which finally underwrote the Society's claims to produce, from its meetings, incontrovertible facts. The word of a gentleman, as a witness, was trustworthy because he was a gentleman. Unlike merchants who might be seen as being concerned with profit, and unlike artisans and scholars who might be held to be in fear and reverence of their masters, the gentleman was held to be completely disinterested, beholden to no one, and therefore entirely unbiased and trustworthy (Shapin, 1988, pp. 395–9; Dear, 1985, pp. 156–7). The special reverence and trust which was accorded to those of gentlemanly status accounts for what might seem to twentieth-century eyes like excessive admiration accorded to Robert Boyle in comparison with a contemporary like Robert Hooke. By modern standards, Hooke seems to have been a better scientist than Boyle. Hooke can be credited, for example, with more discoveries which are still regarded as valid, including, arguably, the so-called Boyle's Law, and yet to their contemporaries Boyle was undoubtedly the greater scientist. Boyle was a gentleman, while Hooke was a paid servant of Boyle's and subsequently a paid employee of the Royal Society. No matter how great the prestige and honour of the Society itself, there was a world of difference, in the eyes of contemporaries, between an economically independent fellow of the Society and one of its paid officials (Shapin, 1989; Pumfrey, 1991).

The gentleman's word, then, was his bond and the testimony of not just two but of 'threescore or an hundred' such disinterested observers was seen as powerful support for the Royal Society's claim to deal only in incontrovertible matters of fact. This provided a ready means of resolving dispute. When the Cambridge theologian, Henry More, tried to use Boyle's air-pump experiments to support his hypothesis of an ubiquitous incorporeal 'Spirit of Nature', Boyle was able to refer back to matters of experimental fact. Boyle's strategy was not to engage in dispute about the existence or otherwise of the 'Spirit of Nature' itself – this would have been to engage in metaphysical dispute – but merely to show that the experimental matters of fact which More explained in his terms could be explained otherwise (remember the metaphor of the hidden clockwork driving the clock's hands), and that other facts simply could not be explained at all by reference to a 'Spirit of Nature' (Henry, 1990). The almost predictable result was that More, being so committed to his own hypothesis, had no choice but to dispute the matter of fact. This was easily dealt with: 'Though he was too civil to give me, *in terminus*, the lye', Boyle wrote, 'yet he did indeed deny the matter of fact to be true. Which I cannot easily think, the experiment having been tried both before our whole society, and very critically, by its royal founder, his majesty himself' (quoted in Shapin, 1988, p. 398). It was More's word against a whole company of gentlemen, including, as Shapin points out, 'the greatest gentleman of all'.

Boyle's and the Society's claim to deal only with matters of fact was extremely shrewd and highly subtle. Most disputes could easily be shown to hinge upon hypothetical issues, like More's 'Spirit of Nature' or Hobbes's conception of the 'simple circular motion' of particles. Once revealed in this light they could be immediately denied any further consideration, thus leaving the realms of scientific, factual, knowledge in a state of undivided harmony. There were occasions, however, as the case of Henry More shows, where matters of fact were disputed. Cases like these had to be very carefully managed if broad claims about the incontrovertibility of factual knowledge were to have any continuing persuasiveness.

Some disputes about facts were dealt with by showing them to be displaced disputes about underlying hypotheses, others by using the overwhelming power of opposed testimony to imply that the disputed observer must be either biased into seeing what he wishes to see (as with More), or too crude and indiscriminating in sensibility to be reliable (as with more 'lowly folk'). Occasionally, however, disputes about facts erupted between rivals who were members of the recognized community of experimental philosophers. In these cases the Royal Society, by the explicit pronouncements of its spokesmen, would act as an adjudicating body whose decision gentlemen were expected to respect. When Johannes Hevelius and Adrien Auzout plotted incompatible paths for a comet which passed through the skies early in 1665, Hevelius referred the dispute to the 'impartial judges' of the Society. Unfortunately for him the Society found in

Auzout's favour. Its secretary, Henry Oldenburg, wrote to him in January 1666:

> Since that time the writings of both of you have been compared and the relevant observations of some of our astronomers examined . . . . These astronomers were unanimous in their opinion that the comet did not by any means draw near to that star in the left ear of the Ram where you seem to have drawn it, without at the same time impugning the phenomenon you saw at that place, whatever that was exactly . . . . And as the observations of the French, the Italians, and the Dutch (in so far as they are known to us) agree wonderfully well with our own, we are quite confident that you will fall in with this consensus of opinion (Oldenburg, 1966, p. 30).

Hevelius's response was two-fold. Firstly, he tried to introduce his own hypothetical considerations into the dispute by offering to explain 'substantial points that I could lay before the public concerning what has been proved and alleged so far' (Oldenburg, 1966, p. 170). In their subsequent correspondence we can see that Oldenburg agrees to have the astronomers of the Society consider Hevelius's points but in February 1667 he is still asking him to be patient in expecting a reply, and eventually the matter is quietly dropped. But Hevelius also responds by taking up the hint in Oldenburg's first reply that Hevelius may have seen something else: 'without at the same time impugning the phenomenon you saw at that place, whatever that was exactly'. In his reply Hevelius suggested that 'the question to be investigated is this: was that phenomenon the self-same comet seen before? You will see that there are very cogent reasons and a variety of observations ready at hand to make this point a very doubtful one and indeed almost ridiculous' (Oldenburg, 1966, p. 172). Oldenburg, concerned to defend the observational competence of his friend as well as to preserve the supposedly unproblematic status of facts, made a number of suggestions, to individual correspondents and in the pages of the *Philosophical Transactions* of the Royal Society, that there must have been two comets. The comet observed by Hevelius could not by then be established as a matter of fact – it had passed across the sky over a year before and no other witnesses had reported seeing it – but its putative existence served to preserve the personal integrity of Hevelius and the continuing validity of the word of a gentleman (Shapin, 1988). It seems clear that for Oldenburg and others in the Royal Society personal honour was an important ingredient in the study of reality.

Boyle and the Royal Society were involved in developing a *convention* for demarcating scientific knowledge. How else can we describe it when it depended crucially on a *local* feature of the social order, that is, the existence of a class of gentry with peer relationships and a distinctive code of honour? Under this system scientific facts could only be reliably formulated by certain people and the conditions of observation are specific conditions of social interaction, as Hobbes made sharply explicit when he pointed out that the Royal Society was not open to all (Hobbes himself was excluded) and its meetings were not conducted in public, but only for the benefit of the fellows (Shapin and Schaffer, 1985, pp. 112–14). And yet, in another sense Boyle's specifications have remained the demarcation criteria of science generally. Modern science depends upon the reports of scientific professionals, much as science, according to Boyle, depended upon 'the word of a gentleman'. There has been a transition in the social basis of fact production: from gentleman to 'virtuoso', to 'philosopher', to 'scientist'. The modern scientist has inherited the special cognitive authority once possessed by the gentleman by virtue of the acknowledged reliability of his word. Modern science is founded on observations in the sense of being validly related to observations made by members of the scientific profession. Boyle's demarcation criterion has changed, yet remained the same: it has evolved.

The nature of Boyle's conventions *as conventions* can be overlooked if one fails to set his writings in their social context. For the specific social arrangements they rely upon do not have to be mentioned in his formulations: he is able to take the social basis of his fact creating system for granted. When we take it for granted too, to the extent of forgetting that it is there, his recommendations take on an appearance of abstract formulations on the basis of which anyone, anywhere might do science. Situated instructions falsely take on the appearance of being unchanging abstract rules.

Might it be said, though, that whilst Boyle's recommendations are indeed conventions, they are (or are close to being) the characteristic conventions for demarcating what we call science, and in that sense are special? Did Boyle get close to describing '*the* scientific method' as the demarcation criterion which accurately grasps what we *today* accept as science? This is not a tenable position for many reasons, not least because philosophers and historians have now demonstrated repeatedly that the contents of the accepted, authentic history of science are not capable of being demarcated by this criterion, or indeed by any other.[2]

It is worth remarking that the Royal Society's demarcation conventions were by no means as clear-cut as Boyle and other leading spokesmen for the Society wished to make out. We have already seen, for example, that although Boyle rejected discussion about divisibility or indivisibility of particles as 'metaphysical' he insisted upon the particulate nature of matter. Similarly, while Boyle dismissed dispute about the possibility of a vacuum as unscientific, he felt no embarrassment about his insistence that air has a 'spring' or a 'restless endeavour to expand itself every way' (quoted from Shapin and Schaffer, 1985, p. 54). The spring of the air, or the corpuscularity of matter might seem to be hypothetical constructs, at least one remove from matters of fact, but for Boyle, various phenomena produced by the air-pump or by chemical experiment served in practice to establish them as matters of fact. Boyle's claims for the factual basis of corpuscularism and the 'spring' of the air did eventually prevail with most of his contemporaries, but there were always difficulties. Some opponents, including the leading philosophers Spinoza and Hobbes, were implacable. Others, such as the leading Continental experimental philosopher Christiaan Huygens, caused specific difficulties.

Huygens noticed what came to be called the 'anomalous suspension' of water. (Here, a barometer tube of water, standing in a trough, is enclosed in the evacuation chamber of an air-pump and a column of water is observed in the tube, even when the air pressure that is held to support it has been removed.) This gave rise to a dispute which threatened to reveal the conventional boundary even of experimental 'facts' themselves. At first, the matter of fact – the suspension of water in the tube when it should have fallen to the same level as that in the trough – was not disputed. The word of Huygens was accepted; his descriptions of his experiments led Boyle, Hooke and others to count themselves as virtual witnesses to the matter of fact. The dispute hinged instead on whether Huygens's pump was sound and hermetically sealed. During the course of this dispute, however, Huygens insisted that Hooke's account of the spring of the air assumed an inherent motion in the particles of air. The 'spring of the air' was in danger of being seen not as an incontrovertible matter of fact, but as a hypothesis dependent on the controversial notion of self-active matter (Shapin and Schaffer, 1985, p. 248). Criticisms of this sort, as Shapin and Schaffer show, had to be very carefully managed. There was a very real danger that the dispute could undermine the role of the air-pump in the experimental programme of the Royal Society. Huygens seemed not only to be threatening the factual status of the spring of the air, but also to be offering experimental proof of the impossibility of a vacuum (since the anomolous suspension of water was explained by Huygens in terms of the supportive pressure of another fluid, more subtle than air which could not be evacuated by the operation of the pump). This was important because it threatened to undermine Boyle's clear distinction between authentic science and supposedly 'metaphysical' issues like the possibility or otherwise of a vacuum. Boyle's strategy for defending the integrity of the air-pump and its role in the experimental programme as a producer of matters of fact, was to show, experimentally, that anomolous suspension (of mercury this time) could be achieved in the open air and so was a phenomenon which had no bearing on what went on within the air-pump (and indeed the phenomenon is now explained in terms of surface tension between the water or mercury and the glass of the tube) (Shapin and Schaffer, 1985, pp. 254–5).

Boyle's and the Royal Society's rhetoric of experimentalism is attractive as an account of what demarcates science in that it is a demarcation which serves to enhance the standing and credibility of science. But there are many such rhetorics. Indeed modern philosophy of science might be read as a catalogue of them. Boyle himself was obliged to face such alternatives, for example, in the polemics of Thomas Hobbes who called into question not only Boyle's 'incontrovertible facts' but his formulated 'scientific method' as well. For Hobbes, so-called matters of fact, deriving as they do from sensory experience, could always be mistaken and should be relegated to the realms of belief and opinion, not knowledge. This was all the more important when dealing with facts which were not directly apprehended in nature, by our unmediated senses, but only as a result of artificial manipulations of machines like the air-pump. Hobbes was perceptive enough to notice that even where one's senses did not go astray, the 'matters of fact' produced by such artefacts always depended upon a set of theoretical assumptions embedded within the actual structure and functioning of the apparatus.

Disputes about beliefs and opinions were inevitable according to Hobbes and guaranteed the failure of Boyle's method. The knowledge required of a philosopher, Hobbes insisted, was the kind of certain knowledge which he took to be afforded by Euclidean geometry. Geometrical knowledge, not knowledge of facts, was indisputable. 'All men by nature reason alike', Hobbes reasoned in contrast to Boyle, 'and well, when they have good principles. For who is so stupid, as both to mistake in geometry, and also to persist in it, when another detects his error to him?' (quoted from Shapin and Schaffer, 1985, p. 100). It is the universal assent that can be commanded by geometry that provides the clue for Hobbes's solution to the problem of order.

Boyle's method was a formulated set of conventions for demarcating scientific knowledge. From a sociological perspective the question is not how good or justifiable it was, or whether some alternative may or may not have been better. Criticism and evaluation of the convention is not appropriate to a sociological perspective. Only empirical curiosity is appropriate. What can be asked from the naturalistic perspective of history and sociology is: why these formulations (Hobbes's and Boyle's) in this context? Shapin and Schaffer ask just this and produce a deeply interesting answer. Let us look quickly at what that answer is.

The crisis of the Restoration settlement made clear just how important it was to find a reliable means of guaranteeing assent. The experience of the Civil War and the Republic had shown that disputed knowledge led to civil strife and public disorder. Both Hobbes and the propagandists of the Royal Society firmly believed that problems of order were problems of knowledge. Any means of guaranteeing assent, even in natural philosophy, could be applied to the political situation to produce social harmony. There was nothing new in this assumption. Throughout the Middle Ages and the Renaissance, natural philosophy had always been regarded as the handmaiden to the Queen of the Sciences, Theology. And theology itself, of course, had always supported the Divine Rights of Kings, and bolstered monarchies all over Europe. The alliance had worked well when there was only one religion and one theology, that of Roman Catholicism. But since the Reformation the system had begun to break down. Different religions with different theologies appropriated rival forms of natural philosophy to act as their handmaidens. The inevitable result was an increasing scepticism towards all systems of knowledge, and a breakdown in all means of assuring assent. By 1660, in England, the consequences of this breakdown in reliable means of knowledge production were all too obvious.

Hobbes, a Royalist appalled by the rebellion against Charles I in 1642, had already made a suggestion for solving the resulting social order problem in his *Leviathan* (1651). At the same time it is a solution to the problem of eliminating dissent and demarcating acceptable knowledge. By showing men what knowledge is, Hobbes was showing them the grounds of assent and so guaranteeing the continuation of social order. The Leviathan, the civil sovereign, enforces both cognitive and social order with the rational support of the individuals of the Commonwealth.

The model of knowledge for Hobbes was geometrical reasoning which could secure total and irrevocable agreement. But even the 'ablest, most attentive, and most practised men, may deceive themselves, and inferre false Conclusions', he pointed out. And so, although Reason itself is a 'certain and infallible Art', 'no one mans Reason, nor the Reason of any one number of men, makes the certaintie; no more than an account is therefore well cast up, because a great many men have unanimously approved it.' Controversies must be settled, Hobbes insisted, by 'the Reason of some Arbitrator, or Judge', who has been appointed by the interested parties acting freely 'by their own accord'. All men reason alike and can be brought alike to see the need to support Leviathan (Hobbes, 1651, pp. 18–9. See Shapin and Schaffer, 1985, p. 323).

The juxtaposition with Hobbes reveals Boyle also to be a political theorist. His doctrines too can be seen as at once solutions to the problems of knowledge, the treatment of dissent, and the maintenance of 'peace'. Unlike Hobbes, Boyle and his supporters believed that it was no longer possible to ensure public uniformity, whether of knowledge or action, by relying upon the guidance of reason. It was the 'unconstant, variable and seductive imposture of Reason' which Walter Charleton, one of the leading fellows of the Royal Society, saw as 'the onely unhappy Cause, to which Religion doth owe all those wide, irreconcileable and numerous rents and schismes' (quoted in Henry, 1992, pp. 196–7). The order problem had still to be solved. The Royal Society ideal, promoted by Boyle, suggested a means of generating and maintaining consensus, of maintaining order without submission to arbitrary authority. Supposedly objective and disinterested, the method of the Royal Society showed a middle way between subjection to authority and radical individualism. Hobbes's method in natural philosophy, as in his political theory, ultimately depended upon unquestioning obedience to an absolute authority. But 'the harsh imposition of uniformity' by a civil sovereign seemed, to Boyle and his colleagues in the Royal Society, as too much like a return to the tyranny of Charles I which had led to the disruptions of the Civil War and the Interregnum, and so Hobbes's Leviathan seemed far from being a viable solution. Peace had to be secured by some other means, in a context where disagreement would undoubtedly continue to exist and to be expressed.

Boyle can be seen as explaining how to create a form of life wherein social stability could be maintained in conditions of less than full agreement. This was the experimental form of life in the first instance. That on which everyone could freely agree was made central to the form of life: namely, the matters of fact evident in nature. That on which disagreement remained was recognized (or rather designated) as peripheral, although still expressible with careful restriction and control. Those who could accept what was central (incontrovertible) and what peripheral (disputable) could enter the public social space of (the Royal) society. Only those who believed in tyranny or in unconstrained individualism needed to be excluded. The model is clearly that, a model of order for the wider context.

As Bruno Latour has said, Boyle's victory over Hobbes stands as a useful symbol of the management of social order in modern society, and Hobbes's defeat symbolizes the passing of an earlier form (Latour, 1993). If Latour is right, Boyle can be seen perhaps as the greatest of all English political theorists. Agree whatever you can and build upon what can be agreed; disagree where you must and expect no resolution; act where you agree and exercise restraint where you do not. This is Sprat's description of the practice of the Royal Society (Wood, 1980; Henry, 1992) and it is Boyle's vision of peace in Restoration England. It is also the continuing recipe for social order in the increasingly differentiated societies of modern Europe.

Conventions could always be otherwise, and how precisely they have been applied could always have been otherwise. Once the demarcation criteria for science are recognized as conventions, this simple truth obviously can be applied to them. Yet many scholars have recognized the conventional character of the various formulations for demarcating science, without taking any interest in why one or another convention is adopted on any given occasion. Probably this is to some extent due to the habit of extending curiosity to conventions of which we disapprove, and somehow treating 'good' conventions as in no need of further understanding. Shapin and Schaffer simply take the step of addressing all contending demarcations of science as phenomena specific to a time and place and in need of explanation as products of that time and space. The results are the magnificent and marvellously well-substantiated insights of their book. They offer a model for the historical and sociological study of boundary drawing in science which deserves to be extended to other places at other times.

6.2 DRAWING BOUNDARIES INSIDE SCIENCE

Just as the boundary separating science from non-science is conventionally drawn so too are those between various specialities within science, and the specific kinds or varieties of knowledge upon which members of these specialities can claim to be authorities. Just as the one convention could be otherwise and is in a sense chosen, so are these others. And just as the choice of the one is a form of political action, so too are the other choices. But in moving from the greater boundary around science to the lesser boundaries within we move from macro- to micro-politics, indeed to something very close to what modern political scientists have identified as bureaucratic politics. Which problems and which phenomena properly belong to which fields is an issue crucially linked to the reputation and standing, the prospects, the resources of all the various fields. And the conventional decisions which have resolved these issues have been the series of historical events which has constituted the institutional structure of science as we now know it with its range of recognized disciplines and specialities.

It is interesting to reflect that in terms of the micro-politics of science every action, however esoteric, however routine, may be said to have a political significance, indeed to be in that sense a political action. Every citation acknowledges a source of authority. Every act of classification extends a particular similarity relation and thereby enlarges the scope of some particular field or speciality rather than another. Every use of an instrument or apparatus testifies to its reliability and hence to the standing of its inventors or current guardians. Scientists do not engage in the 'social' activity of making boundaries, and the 'technical' activity of doing science within these boundaries as two qualitatively different kinds of action. For the most part they create and sustain boundaries in the course of, as part of the business of, doing science. Only in periods of controversy and disputation do we sometimes see the micro-politics of science separating off into a sphere of reflective debate separate from practical activity. And even here the differentiation of the 'social' and the 'technical' is temporary and incomplete.

The contrast of the macro- and micro-politics of science may be misleading, however. Detail work in science is never intelligible purely by reference to the esoteric conventions and concerns of the speciality in which it is performed. It always has significance for allied or opposed specialities, and is always liable to evaluation as an element of science generally and an instance of what is conventionally accepted as science. Shapin and Schaffer show how ideals of political order can account for different opinions of the permeability of a leather washer in the seal of an air-pump. There is no natural divide between large and small matters in the practice of science, or between major problems and minutiae, nor is there a natural divide between macro- and micro-politics.

An example which illuminates most of these points is provided by the reception of Robert Chambers's *Vestiges of the Natural History of Creation*, published anonymously in 1844. The book was to draw its author into a number of highly technical and esoteric debates with a range of specialized 'scientists', whilst at the same time engendering debates on what was and what was not 'genuine science' and whether *Vestiges* in particular deserved to be so considered. We shall take the reception of this book as the case study with which to illustrate the process of boundary drawing within science, just as we took Boyle's dispute with Hobbes as the vehicle for the previous discussion.

All the leading scientist reviewers of Victorian Britain vilified *Vestiges*, and ridiculed most of its particular claims. But it nonetheless became a bestseller and, although it 'never had a single declared adherent', as the author himself wrote (Chambers, 1853, p. ix), it must now be seen as one of the most influential works in the development of Victorian science. The *Vestiges* brought together a number of current ideas from different emergent scientific sub-disciplines to elucidate what Victorian naturalists called 'The Mystery of Mysteries', the problem of the origin of species. The *Vestiges* was the first systematic attempt by a British writer to establish a theory of biological evolution (predating the second attempt, Darwin's *Origin of Species*, by fifteen years).

The book actually attempted to establish evolution as a *universal* process under the direction of natural law, a law of progress. Beginning with a discussion of the nebular hypothesis, in which the solar system was held to have evolved by precipitation out of a cloud of hot luminous matter, the argument proceeded by drawing upon the latest ideas in geology, paleontology and various biological fields, including phrenology, spontaneous generation, comparative anatomy and comparative embryology. By appropriating ideas from these different areas, all of which were currently being debated in scientific circles (even if some of them, like phrenology and spontaneous generation, were controversial), the author, Robert Chambers, was able to point to apparently overwhelming evidence of a trend from chaos to order, from simplicity to complexity, from primitive to more advanced states of affairs.

In some cases Chambers simply adopted the standard consensus of contemporary 'scientists', in other cases he adapted current ideas by interpreting them in accordance with his evolutionary views. The nebular hypothesis was taken directly from the writings of William Herschel, Sir David Brewster and John Stuart Mill, all distinguished Victorian intellectuals. The progression of life forms in the paleontological record, from cephalopods, to fishes, reptiles, then mammals and finally 'man', was the standard view amongst geologists. Chambers put a highly unorthodox evolutionary gloss on this progressionism, however, supposing that earlier forms had *evolved* into the later ones. For his evidence he took Richard Owen's theory of homologies, developed to explain why there should be a one-to-one correlation of the bones in the forelimbs of a mole with those in the flippers of a dugong and the wings of a bat, and the embryological studies of F. Tiedemann and Étienne Serres who claimed that the brain of a mammal foetus develops through stages in which it first of all resembles the foetal brain of a fish, then of a bird, a reptile and finally takes on the distinctive features of mammal brains. The theories of Owen and Tiedemann and Serres were not originally presented as evolutionary theories but Chambers found it easy to adapt them to his purpose.

The uniformly negative reaction of 'scientists' to the *Vestiges* presents us with a historical problem. We need to ask why it provoked this kind of response even though every aspect of Chambers's thesis was taken from the current repertoire of science. None of the ideas was new and all had respectable adherents in accepted scientific fields. Moreover, the Victorian reading public made the *Vestiges* a bestseller, so it could scarcely be said to be detracting from the public standing and credibility of science. Chambers's theory of evolution could reinforce, better perhaps than any other single work, the increasingly popular notion that the prevailing *laissez-faire* political economy, with its concomitant morality of the devil take the hindmost, was entirely *natural*, that it was as much a part of God's plan as the extermination of species. If the educated public liked the *Vestiges* why didn't the 'scientists'? A large part of the answer lies in the fact that the *Vestiges* was seen by members of this rapidly professionalizing occupation as a major hindrance to current efforts to demarcate boundaries, not only between science and non-science but also between different disciplines within science (Yeo, 1984). Chambers himself was convinced that the real reason for the scientific criticism of *Vestiges* was that its author was seen to be an amateur who had no right to speak of science. In a sequel published in 1845, entitled *Explanations*, Chambers tried to answer all the objections raised against the *Vestiges*. Having dealt with all the specific criticisms he pointed out that there was one final complaint 'likely to be of some avail with many minds', and that was that the *Vestiges* must be wrong because 'nearly all the scientific men are opposed to the theory'. Chambers was unabashed:

If I did not think there were reasons independent of judgment for the scientific class coming so generally to this conclusion, I might feel the more embarrassed in presenting myself in direct opposition to so many men possessing talents and information. As the case really stands, the ability of this class to give at the present time a true response upon such a subject, appears extremely challengeable. (Chambers, 1994b, p. 175)

Chambers went on to identify what it was 'independent of [scientific] judgement' ·which prevented the 'scientific class' from giving a 'true response' to the *Vestiges*: it was the increased specialization demanded by the move towards professionalization in science. As he put it: 'Experiments in however narrow a walk, facts of whatever minuteness, make reputations in scientific societies; all beyond is regarded with suspicion and distrust' (p. 176). The result, therefore, is that when the layman requires the answer to really far-reaching and important questions,

the man of science turns to his collection of shells or butterflies, to his electric machine or his retort, and is mute as a child who, sporting on the beach, is asked what lands lie beyond the great ocean which stretches before him. The natural sense of men who do not happen to have taken a taste for the coleoptera or the laws of fluids, revolts at the sterility of such pursuits.

Eventually Chambers came to the conclusion that, where specialism and a refusal to speak on topics outside of specialism reigned, the *Vestiges* would never get a fair hearing. Accordingly, he wrote, 'it must be before another tribunal, that this new philosophy is to be truly and righteously judged' (p. 179). The professional interests of the scientists, in other words, made them inadequate judges of the truth of *Vestiges*. The correct tribunal was to be made up of intelligent laymen. Chambers's concerns here echo those of Hobbes in his denial of the validity of Royal Society science. Hobbes himself was denied entry to the ranks of Royal Society fellows and he accordingly took a sceptical view of their claims to present matters of fact which were 'publicly' witnessed. The situation in science seemed, to outsiders like Chambers and Hobbes, to be much closer to that of other professional groups, creating and sustaining 'knowledge' claims for ideological purposes, in particular to serve their own social interests and to provide themselves with illegitimate claims to authority (Shapin and Schaffer, 1985, pp. 92, 333). For Chambers the narrow specialisms of science were obstacles to progress. The scientists' insistence on demarcating boundaries around narrow specialisms and refusal to enter into discussion of the wider implications of current scientific theories made it impossible for developmentalism in biology to emerge as a viable scientific theory. But the ideological developmentalism of the Victorian middle class was precisely what *Vestiges* was promoting, Chambers believed in the politics of self-improvement and *laissez-faire* (Secord, 1989).

Needless to say, the scientists were recalcitrant. Hugh Miller, who wrote an extended critique of the *Vestiges* based on the paleontological evidence provided by a single fossil, tried to legitimate the highly localized boundaries of the Victorian scientists. Surely, he insisted, the specialist scientist has a better right to be heard on his own subject than 'any other class of persons' (Miller, 1870, p. 264):

> We must surely not refuse to the man of science what we at once grant to the common mechanic. A cotton-weaver or calico printer may be a very narrow man, 'exclusively engaged in his own little department'; and yet certain it is that, in a question of cotton-weaving or calico-printing, his evidence is justly deemed more conclusive in courts of law than that of any other man, however much his superior in general breadth and intelligence.

Persuasive as Miller's rhetoric seems, it is hard to reconcile with what we know of the actual treatment of the *Vestiges*. According to Miller, the author of the *Vestiges* calls upon astronomy, geology, phytology and zoology in such a way that the astronomer objects to the astronomy, the geologist to the geology and so on. What actually seems to have occurred is that the astronomers objected to the geology, the geologists objected to the astronomy, the zoologists objected to the geology. They were forced to do this because Chambers was always careful to draw the different strands of his argument from the leading authorities in the appropriate branch of science. He succeeded so well in this that Richard Owen, Britain's leading comparative anatomist and paleontologist, acknowledged in a letter to the author of *Vestiges* that the 'few mistakes where you treat of my own department of science' can be 'easily rectified in your second edition' (quoted from Yeo, 1984, p. 12). The critical reviewer for the *Athenaeum* conceded that the author had 'produced a book which may in some manner serve as an outline to the vast range of the natural sciences' (quoted from Yeo, 1984, p. 12). And in general it was Chambers's impressive grasp of the latest scientific ideas, and the implications they had for one another, that made the *Vestiges* so difficult to refute. In those cases where, say, a geologist did object to his geology, Chambers was more often than not able to answer the objection. Consider, for example, Chambers's devastating response to various geological claims to have found fossil fish from an earlier period than hitherto thought possible, and so early as to refute the developmental hypothesis. Chambers added a note to the tenth edition of his *Vestiges* (1853) in which he wrote:

> From 'a very competent authority', we learn that the seven species of fish from the Wenlock Limestone, believed in by Professor Sedgwick, because he had 'seen' them, turn out to have been found in the debris of a quarry of that rock, where it is admitted they had most probably been dropped from the pocket of a workman who had obtained them in a neighbouring quarry of a higher formation! The Onchus spine from the Bala Limestone had been entered by the government surveyors, as a fragment of fish, after only a 'cursory examination', it proves to be, 'in reality, half the rostral shield of a trilobite!' 'Its resemblance to an Onchus was due merely to its being broken in half and obscured by stone.' In like manner, the spine from the Llandeillo flags, certified as such by 'one of the most cautious and practised geologists of the present age', has been declared to be nothing but a piece of 'a new genus of Asteroid Zoophyte', something lower in creation than even the Bala spine proved to be! (Chambers, 1853, p. xxxvi)

Once again, it should be noted, Chambers is drawing his information from 'a very competent authority', one who was, unlike Chambers, judged by his peers to be an accredited geologist. The source was J.W. Salter writing in the *Journal of the Geological Society*, May, 1851.

More frequently, however, critics of the *Vestiges* strayed, in desperation, outside their own field of expertise. Miller, a geologist, tried to reject the botanical theory of Lamarck and Lorenz Oken that land plants developed from seaweeds by insisting that there are no intermediate kinds of plants found on shorelines or in estuaries where there is a gradual transition from salt to fresh water. Chambers questioned the theorizing and, more significantly, the facts, in terms full of ironical import for those concerned by his scorn of disciplinary boundaries. 'I must confess', he wrote, 'that till botanists of good authority have given their opinions, I cannot consent to take Mr Miller's statements as conclusive even about the facts' (Chambers, 1853, p. xliv).

Similarly, in dismissing the claims of Adam Sedgwick, professor of geology at Cambridge, that the development of embryos does not recapitulate the progression of life forms found in the geological record, Chambers was able to refer to one of the world's leading zoologists, Louis Agassiz. Sedgwick was 'neither an anatomist nor a naturalist', Chambers emphasized, and so his words cannot be allowed to have any force in embryology over those of Agassiz. Chambers even drew attention to the lapses and inconsistencies of expert scientists as they policed the boundaries of their own subjects. While Chambers is pulled up for tiny infringements, 'a mere working geologist and litterateur like Mr Miller' is allowed to get away with far worse sins. After drawing once again upon the work of Agassiz to point out an error in Miller's claims about a fossil fish, Chambers wrote:

> It is perhaps no great discredit to Mr Miller who is not a naturalist, to have proceeded on such false grounds; but it is somewhat remarkable that naturalists should have allowed his work to be four times printed, without giving a hint for the warning of his readers, while the most trivial mis-statements of the present work, albeit in general not at all bearing on its conclusions, have been held up as destructive of its credit.

The selective use of scientific argument here derives from the differently perceived *social statuses* of the author of the *Vestiges* and of Miller. Miller was recognized to be a genuine researcher in his chosen field whereas 'the Vestigiarian' was merely a compiler of facts and theories that had been produced by others. As T.H. Huxley put it, the auth · of the *Vestiges* seemed to him to be someone 'who would have been an astronomer – but for sitting up at night; a geologist – but for dirtying his fingers' (Huxley, 1854, p. 438). Whoever the author of the *Vestiges* was (Chambers always hid behind a screen of anonymity), he clearly didn't count. He had not earned the right to speculate by appropriate toil in *any* scientific field, much less in *all* of them. Speculation could only be entrusted to those who had paid their dues by doing detailed research, in a properly delineated special field of professional science.

The *Vestiges* drew heavy and prolonged censure from contemporary scientists because it was perceived to be a threat to the image of science as the preserve of specialized experts. An expert in science, or rather in a particular branch of science – since by this time it was already agreed that no single individual could be the master of science in its entirety – was one who had proved himself to be a reliable authority by dint of empirical research and theorizing of an appropriately cautious kind. But the expert was also expected to acknowledge the limits of his own expertise and defer to the expertise of others in adjacent fields. This served the general professional interest of scientists: the credibility of scientists was being linked with their being specialized practitioners. The author of the *Vestiges* did nothing to help this enterprise, and seemed rather to subvert it. To admit the value of the *Vestiges* was to admit that profound truths were accessible without specialization, without extended training, without narrowing of vision and purpose, in a nutshell, without science.

The *Vestiges* was, however, more than a threat to *generalized* efforts to promote a particular image of science. At the same time it directly affected the professional consensus – or the attempted consensus – as to what was to count as genuine 'scientific knowledge' and what was not. And it endangered the boundaries which had been carefully erected between specific sub-disciplines within the sciences.

The evolutionism of the *Vestiges* directly threatened to undermine the prevailing principles of geological science and the newly established relationship between geology and biology. We have already seen that the consensus among most geologists was that there was a progression of life forms in the rocks. The fossil evidence of the development of life seemed to fit in well with the observational evidence of the physical development of the earth itself, and geological theory had been developed to underwrite both sets of evidence. The theoretical assumption of the geologists was that the earth had begun its career as a molten ball of matter and that it was still undergoing gradual cooling (this was demanded by contemporary thermodynamics, and was an extension of the Laplaceian nebular hypothesis). It seemed to be an obvious corollary of this theory that in the early stages of cooling the earth's crust, forming on the outside, would be subjected to greater forces and stresses than those which now affect it, given that it has become much cooler and, therefore, more stable. Furthermore, the geological record showed ample confirmation that the rocks of the earth had been subjected to immense forces, thrusting up mountain chains, rotating whole beds of strata through 180°, and so on. The earlier history of the earth, then, seemed to be a tale of cataclysmic events, sudden disruptions of the order of nature which distorted the rocks and caused tidal waves and flooding of such force that they brought about massive erosion effects in an extremely short period of time. This was catastrophist geology.

It is true that by the 1830s there was an alternative, and rival, view within geology which held that all the manifestations of earlier disturbances, erosion, destruction and elevation, were brought about by very gradual means, by precisely the same, almost invisibly slow earth-forming processes that can be detected all around us today. But this so-called uniformitarian view of geology was generally held to be incompatible with two important pieces of evidence. The first of these was the thermodynamic estimates of the maximum length of time for the age of the earth made by physicists. The second was the paleontological evidence.

The fossil record showed that the characteristic flora and fauna of a given period in the past, as judged by the rocks in which their fossil remains are found, were overlaid in subsequent rocks by a completely new set of flora and fauna. As the majority of geologists read this, one population had been wiped out suddenly and replaced by a new set of different life forms. The complete disjunction between the successive sets of creatures seemed to be inexplicable without assuming that the earlier forms had been overtaken by some very sudden, cataclysmic event. Given that this view seemed to imply short time-scales – compared with the uniformitarian view – and therefore sat more easily with the time-scales suggested by contemporary thermodynamic theories, most geologists embraced catastrophism.

Any suggestion, therefore, that the life forms of one geological period had gradually evolved into the forms that were characteristic of a subsequent period seemed to be incompatible with the precepts of catastrophist geology. Gradual evolution was inconsistent with the notion that the earth's history was punctuated by massive, life-annihilating cataclysms, and could be accepted only if the apparently obvious and unambiguous observational evidence of paleontology was disregarded. Evolution was dismissed as unscientific accordingly.

In making this argument, however, the geologists were drawing a particular boundary around their science. Geologists could properly give detailed accounts of how the successive layers of rock had been laid down at different ages, and what changes they had undergone since, and when and why. Moreover, the fossil record could be used to provide extra circumstantial evidence for these alleged developments. What the fossil record could not do, however, and what no other kind of geological evidence could do, was to reveal just how the new populations of flora and fauna, which emerged *after* a supposed geological cataclysm, actually came into being. The problem of the origin of species, accordingly, was regarded as being beyond the legitimate realms of science. On this issue, William Whewell wrote, 'men of real science do not venture to return an answer' (Whewell, 1846, p. 21).

By erecting a boundary around the cognitive issue of what can and cannot legitimately be said about the fossil record, the geologists were bound to attract the attention of biologists. Fossils were, to be sure, rock formations but they were also acknowledged to be records of plant and animal forms. Biology was a comparatively new science and as such was busily engaged in trying to define its own boundaries. This being so, there was inevitably some friction over the issues raised by paleontology. In a superb study, the late Dov Ospovat pointed to two separate traditions in what was then called 'the History of Life' (1981). He designated them 'teleological' and 'developmentalist' respectively. Teleological explanations routinely assumed that plant and animal forms were designed by God with the specific purpose of filling what we would now call 'an ecological niche'. The underlying assumption was that the succession of organic forms on the earth has been wholly dependent on the development of the earth itself, even if the precise manner of formation of these new organisms remained unknown (and as Whewell would have it, scientifically speaking, unknowable). This was essentially the view held by the overwhelming majority of geologists.

The alternative view eschewed teleological explanation as unscientific and insisted that it should be possible to find biological forces or laws (such as the law of unity of type, or the law of succession of the same type in the same country) which explained the successive development of life forms in exactly the same way as geological forces and laws explained the changing nature of the habitat. This was the line taken by the majority of would-be biologists, when the domain of biology itself was only just being staked out.[3] For such men, the geologists were conceding too much to religion. In spite of the seeming methodological impossibility of learning anything about the actual origin of any given species, the biologists believed that it should still be possible to discover biological laws which determined the development of living creatures. These laws would be the equivalent of the laws of motion in physics, laws which said nothing about why or how motion came to exist, but which described and explained all specific examples of bodies in motion.

Among the biologists' first efforts to develop a philosophical, or lawlike, account of zoological and botanical phenomena was the attempt to explain what Richard Owen called homologies, and the phenomena of generation, growth and embryological development. It would have counted as a failure of biology to prove itself a science if the phenomena it purported to deal with could only be explained by reference to the direct action of a divine creator. This would have been to ally biology with religion, rather than to situate it in the realm of the sciences. Biologists like Carpenter and Owen (as well as a number of Continental biologists) were delighted, therefore, to discern recurring patterns in life forms. This suggested the existence of biological laws through which the creator chose to work, in the same way that he chose to create the solar system and the earth itself, not directly, but through the inexorable action of the laws of motion and the universal principle of gravitation.[4] There were, biologists claimed, various archetypes of living forms which developed in certain ways, according to still undiscovered laws of development. But the archetypal forms and the laws governing their development could be traced to an even deeper level. The development of the embryo mammal was seen as a recapitulation of the progressive development of mammals from the older vertebrates, as revealed by the fossil record. The conclusion once again was that God had brought about the variety of living creatures not directly but by establishing scientific laws to govern the development of a certain limited number of archetypes. The self-appointed role of the biologist was to discover these laws.

Both geologists and biologists, then, believed in a temporal progression of life forms from the less complex to the more complex, but each group made use of this supposed knowledge in a different way. Geologists took the progression of life as a taken-for-granted matter of fact, confirmed by the expertise of comparative anatomists, which (because of the radical discontinuities between successive forms) gave further support to their catastrophist geological theories. While the biologists set themselves the task of elucidating the laws which governed the progressive nature of successive animal and plant forms.

The border patrols on both sides of the boundary couldn't help feeling uneasy, however. Developmentalism in biology was obviously compatible with evolutionism (which is, indeed, a special case of developmentalism) but the majority of scientists at the time, biologists as well as geologists, saw the need to deny the scientific legitimacy of theories of biological evolution. As Ospovat points out, the biologists were 'in a delicate situation': 'the issue of biological explanation was a sensitive one' (p. 19). The biologists needed to explain the historical

progression of life forms as revealed in the rocks, the similarity of complex creatures to simpler ones during their embryological development,
and the homologies between the organs of widely different species,
without recourse to evolutionary explanations. For most biologists
evolutionary theories were already excluded from their considerations,
having being designated as 'unscientific' by natural theologians and by
catastrophist geologists alike.[5]

As a result, the biologists did indeed tread very carefully: anti-
evolutionary developmentalism was shown to be compatible with
geological catastrophism (and, incidentally, with natural theology)
while evolutionary forms of developmentalism were rejected and
denied. It is as though the agenda had been set by the older science
of geology, and the biologists, struggling to erect agreed boundaries
around their own sub-discipline, tried to show how their work could
be seen as 'independent' confirmation of the work of the geologists.
It is not surprising, therefore, that we find the further researches
of biologists presented in such a way as to deny the possibility
of evolution. All the evidence points to the fact that species are
fixed, they insisted.[6] Had the biologists tried to make evolutionary
theories a part of their sub-discipline, they would have been branded
by geologists, natural theologians, thermodynamic physicists and others
as pedlars of pseudoscience. The origin of species for both geologists
and biologists was regarded as beyond the legitimate realms of science.
Biologists demarcated their own area of expertise without transgressing
or disrupting the boundaries erected by other scientists. On the contrary,
the biologists reinforced the geologists' exclusion of evolutionists by
shoring up the wall from their side too. As Robert Frost wrote in his
poem 'Mending Wall', 'good fences make good neighbours'.[7]

It is easy to see that the *Vestiges* transgressed the boundaries carefully
erected between the would-be good neighbours of geology and biology.
Indeed, it seems likely that one of the major indications to contemporary
scientists that the anonymous author of the *Vestiges* was an amateur was
the very fact that he did not respect professional boundaries. As David
Brewster condescendingly suggested in a review, the author should
concentrate on one 'little department of science' before he should
try to generalize. While another critic suggested that if the author
of the *Vestiges* had 'performed one single chemical experiment, and
endeavoured to understand its import' he would have understood the
interconnectedness of scientific ideas and 'would never have presumed
to write this book' (quoted in Yeo, 1984, p. 24). Chambers's failure to

respect borders probably also explains why his critics were so often drawn across those borders themselves. A geologist like Sedgwick or Miller, used to being able to proceed without reference to the latest work in comparative embryology, and fully confident that, where such work did relate to their own theories, it would appear in a supportive role, were no doubt much taken aback to find such work presented in a threatening way. Feeling that this use of comparative embryology (or whatever else it might be) could not safely be ignored, they could only meet it head on.

We saw above how Chambers called for the truth of his theories to be judged not by narrow specialists but by intelligent laymen. Significantly, the leading spokesman on behalf of science in Victorian Britain, William Whewell, responded to the *Vestiges* and its popularity in precisely the opposite way. 'If the mere combining chemistry, geology, physiology and the like, into a nominal system, while you violate the principles of each at every step of your hypothesis, be held to be a philosophical merit', he wrote, then 'I do not see what we . . . have to do, except seek another audience' (quoted in Yeo, 1984, p. 26). Scientists, in other words, should forget the lay public, stay within their own boundaries and talk only to their fellow scientists.

It is perhaps worth remarking here that when Charles Darwin finally published his own version of evolutionary biology in 1859, after keeping it essentially private for many years, it led very quickly to the acceptance of evolutionism (but not Darwinism) by the majority of biologists. Darwin, unlike Chambers, had the highest possible credentials as a 'man of science', and was known in geological as well as biological circles, and this made all the difference to the reception of his work. Even so, the transfer of evolutionism into mainstream biology from its former position beyond the pale had to be very carefully managed. The story is complex and we cannot enter into it here but it should come as no surprise that the boundaries of geology were completely renegotiated. Catastrophism in geology, with its emphasis on the discontinuities and dramatic changes in the fossil record, gave way to so-called uniformitarian theories which asserted the reality of gradual change over long periods of time.[8]

In conclusion, then, we can see that Chambers's anonymous book was denounced by the leading spokesmen of science in Victorian Britain because it undermined various boundary drawing enterprises which the overwhelming majority of scientists saw as crucial. Both this, and the earlier case study, illustrate how the boundaries of science stand as the outcome of the strategic activities of scientists themselves. The boundaries of science are conventional. To reify those boundaries, and to see them as hard-and-fast divisions between inherently different subject areas or disciplines, is simply a mistake.[9] The demarcation of science from pseudo-science, or of science from scientism or even physics from chemistry, can be fully understood only in sociological terms. Any future attempt to draw fixed demarcation criteria must be greeted with a question: why has this attempt been made? This can only be answered by looking at the social and historical context. Scientific boundaries are defined and maintained by social groups concerned to protect and promote their cognitive authority, intellectual hegemony, professional integrity, and whatever political and economic power they might be able to command by attaining these things.[10]

Not surprisingly, since science is one of our most successful and securely established institutions, its boundaries have become deeply entrenched. Deeply entrenched conventions are easy to mistake for natural divisions and there are strong incentives which encourage such mistakes. Mistakes, however, they remain, and potentially expensive ones at that.

CHAPTER 7

Proof and Self-Evidence

So far we have discussed empirical knowledge. There exists, however, a long tradition in which knowledge is deemed to fall into two kinds: empirical knowledge and a species of non-empirical cognition represented by mathematics and logic. Typically the contrast is drawn by speaking of contingent *versus* necessary truths, and sometimes different mental faculties are cited as the sources of the knowledge, e.g. sense experience gives us knowledge of contingent matters of empirical fact, while the faculty of reason allows us to grasp the necessary truths of logic and mathematics. In this final chapter we will argue that the approaches and results that have been discussed so far can be extended from the realm of the empirical and contingent into the realm of 'necessary' truth.

The most obvious difficulty facing the sociologist of mathematics lies in coming to grips with the technical content of the subject under study. No one could exhibit, say, the role played by competing interests in the disputes over the measure of statistical correlation, without being able to follow the mathematical reasoning involved (cf. MacKenzie, 1981). This can be a formidable obstacle for the sociologist, just as it is for the historian or philosopher of mathematics. For the sociologist there is a further and more subtle difficulty: the problem of finding the right perspective with which to approach the technical content. Without such a perspective, some of the most stubborn facts about mathematics, the ones that seem maximally resistant to sociological analysis, appear to be amongst the most trivial, e.g. the equation telling us that $2 + 2 = 4$.

This equation has, indeed, assumed an ideological significance. It has become a symbol of the absence of social interference with the ideal of free and rational thought. In Orwell's *Nineteen Eighty-Four* (1949) O'Brian, the sinister representative of totalitarianism, asks Winston Smith if he recalls writing in his diary, 'Freedom is the freedom to say that two plus two makes four'? He then confronts Smith with the four extended fingers of his left hand, and asks him how many there are. Given the answer 'Four', he responds: 'And if the Party says it is not four but five – then how many?' (p. 200). Smith is tortured and brainwashed. Eventually he comes to believe the Party slogans, such as 'Freedom is Slavery' and, significantly, 'Two and Two make Five' (p. 223).

It is perhaps not surprising that hostile critics of the sociology of knowledge have fastened upon this equation and used it as the basis of their opposition. Social conventions, or more extravagant forms of social coercion, might be needed to explain why someone doesn't believe that $2 + 2 = 4$, but when they do believe this evident truth there is surely nothing significant for the sociologist to say. This is why we have chosen in this chapter to address the question: what can the sociologist of knowledge say about $2 + 2 = 4$? We will begin by looking at a typical formulation of the challenge it is taken to pose.

7.2 THE CHALLENGE

In his book *Progress and its Problems* Larry Laudan says:

> There is an enormous amount of evidence which shows that certain doctrines and ideas bear no straightforward relation to the exigencies of social circumstances: to cite but two examples, the principle that '$2 + 2 = 4$' or the idea that 'most heavy bodies fall downwards when released' are beliefs to which persons from a wide variety of cultural and social situations subscribe. Anyone who would suggest that such beliefs are socially determined or conditioned would betray remarkable ignorance of the ways in which such beliefs are generated and established. (Laudan, 1977, p. 200)

There are four significant points here. First, the critic fastens on what he calls the principle, doctrine or idea that $2 + 2 = 4$. Second, he says that its acceptance bears no straightforward relationship to social circumstances. Third, he says that the belief that $2 + 2 = 4$ is subscribed to by people in a variety of cultural and social situations. Fourth, he claims that the way this belief is generated and established is sufficient to show that it isn't socially determined or conditioned. Let us take each point in turn.

First, what exactly is the belief that is being offered as a counter-example to the sociology of knowledge? We should notice that the meaning of symbols such as $2 + 2 = 4$ is variable and context dependent. Consider the following. Someone is asked if they accept that $2 + 2 = 4$? On agreeing to this, they are asked to add $3 + 3$, and they give the answer

1. Asked to add 3 + 4, they give the answer 2. Asked to multiply 2 × 2 they say 4; but when asked to multiply 3 × 3, they also give the answer 4. We are imagining, of course, someone who is working within a finite arithmetic whose five elements are the numbers 0, 1, 2, 3, 4. This means, in effect, that its users will only be interested in the remainders that are left after dividing through by five. The number of fives involved is then, as it were, discarded; hence both 2 × 2 and 3 × 3 are equal to 4. This is an acceptable and interesting mathematical system. It has just been described in a way that makes it seem parasitical on what we may call our 'ordinary arithmetic', but that was merely a matter of exegesis. The system can stand entirely on its own, with its own set of rules.

We must be careful when we speak of 'the' principle, or 'the' idea that 2 + 2 = 4. Familiar symbols can be located in a variety of different systems which obey disconcertingly different rules. We are not suggesting that the author of the critical passage is unaware of this, but the mathematical simplification leads to a sociological simplification. Many people would be deeply puzzled to hear it seriously asserted that 3 + 3 = 1 and 3 + 4 = 2, alongside 2 + 2 = 4. To appreciate these assertions as anything other than error and nonsense requires exposure to mathematical culture. Such exposure is neither intense nor uniform in our society. What the critic called 'the exigencies of social circumstances' – in this case, the contingencies of our mathematical education – are therefore not so irrelevant as he would have us believe. We will come back to this later, as we will to the emotionally charged case of 2 + 2 = 5.

The second and third points can be dealt with more quickly. We are told that accepting 2 + 2 = 4 bears no straightforward relationship to social circumstances. This may be true, but it cannot be made to bear much weight. Sociological explanation does not require that specific mathematical beliefs (e.g. Pythagoras' theorem) be tied to some special social context (e.g. the Greek city state). An explanation would be no less social if it demonstrated a revealing connection between such a belief and a recurrent social feature that was common to many otherwise different contexts; or if it could be shown how a practice was sustained because it served a useful purpose for agents in diverse circumstances. A widespread belief can be explained by a widespread cause, or by a variety of different causes which all tend to the same effect. In the case of 2 + 2 = 4, psychological and sociological considerations would lead us to expect that such an equation might have widespread acceptance. Its utility for organizing practical affairs is evident, e.g. its role in keeping track of people in a group, counting cattle and conducting primitive barter. Such universality as it enjoys may be intelligible, given these basic and diverse social concerns.

The first three of the points in the critical passage therefore pose no problem of principle. What of the fourth and last point?

7.3 THE NUB OF THE ARGUMENT

The critic appeals to the way our knowledge of 2 + 2 = 4 is 'generated and established'. He says that only 'remarkable ignorance' of this could lead anyone to countenance the idea that such knowledge is socially determined or conditioned. It would have been helpful to hear the critic's own account of how 2 + 2 = 4 is generated and established. This would permit a direct comparison with any candidate sociological analysis. Unfortunately no such account is provided. So what does the critic mean by the 'generation' of a belief? Does it refer to its psychological and social origins and its emergence within the individual mind? This would involve a study of whatever educational processes attended the formation of the belief. We can be confident that this was *not* the critic's intention, because these processes do not stand in sharp contrast to social determination and conditioning. The training to which children are subjected by their teachers as they learn arithmetic involves the very things the critic wants to avoid. What could be a clearer resort to authority than the insistence that in learning to count the child *has* to say 5 after 4, and 4 after 3? We are drilled in counting, just as we are drilled in the alphabet; we learn to add as we learn to spell: and yet no one treats alphabets or the rules of spelling as anything more than social conventions.

We conclude that the critic must have had something other than socialization in mind when he used the word 'generate'. The accompanying use of the word 'establish' gives us a clue. As well as its psychological and social meaning it can also mean 'prove'. To establish that 2 + 2 = 4 can mean to prove that 2 + 2 = 4. 'Generation' would then be a reference to the way the theorem that 2 + 2 = 4 emerges from a proof. We shall therefore proceed on the assumption that the challenge reduces to this: how can we think that a sociological account is called for, given that 2 + 2 = 4 can be *proven*? For our critics the credibility of 2 + 2 = 4 is a consequence of rational conviction generated by proofs. We shall recommend the view that 2 + 2 = 4 is not credited on the basis of proof and, further, that proofs with the requisite compelling character are not to be found. It will turn out that we do not believe 2 + 2 = 4 because we are compelled by a proof, but that we are compelled by the proof because we already believe 2 + 2 = 4.

The ground rules for the subsequent discussion can now be laid out. First, any discussion of the sociology of knowledge that appeals to proofs must centrally concern itself with the people who devise and use the proof, e.g. with the interaction of the proof and prover. It is what proofs do for our state of knowledge that matters. Second, if the sociology of knowledge is to generate empirically grounded, scientific claims of its own, its approach must be naturalistic. We must therefore view with suspicion any appeal to alleged ways of knowing mathematical truths that are inherently mysterious, e.g. dependent on supersensible forms of intuition or the direct, rational apprehension of such truths through non-causal channels. The challenge is to see if a naturalistic approach (combining psychology and sociology) can do justice to the role played by 2 + 2 = 4 as an established theorem of arithmetic.[1] Now back to the question of proof.

7.4 THE NAIVE PROOF

What counts as a proof that 2 + 2 = 4 varies markedly in different segments of our culture. Just as the knowledge of finite arithmetics is socially distributed, so is the understanding of the procedures necessary to establish that 2 + 2 = 4. Two extreme styles of proof can be distinguished. They may be called 'low-status' and 'high-status' respectively. Low-status proofs are the naive procedures that would satisfy the lay person. Elementary teachers might produce such a proof in a school classroom. High-status proofs by contrast would only be offered by professional logicians. Let us work our way up the hierarchy, starting from the bottom.

A mathematically unsophisticated person might find it strange to be asked to *prove* that 2 + 2 = 4, but if pressed would proceed as follows. Starting with a group of objects – assume they are apples – they would select a pair of them and place them in a prominent position, saying 'Look – one, two'. They would repeat the performance with another pair, and then conspicuously bring the two pairs together. Then they would count them: 'one, two, three, four'. The claim might then be made that all similar groups of things would produce the same result. That is why 2 + 2 = 4.

Those higher in the cultural hierarchy would deny that this is a proof – a reaction that would probably be met with perplexity from below. What more could be done? The idea that anything further might be required depends on a special way of looking at things: it depends on a strong sense of the difference between inductive and deductive procedures. Given this distinction we can draw attention to the fact that just because *these* apples behave like this *now*, that doesn't show that they – or anything else – will *always* respond in that way. Consequently, all that the so-called proof has done is to verify an instance where two of these things plus two more of these things makes four of these things, and then generalize the result. This turns the arithmetical proposition into something like a law of nature, an inductive and empirical truth as best. And that surely is different from establishing the timeless truth that 2 + 2 always and *must* equal 4. We may express this by saying that the shortcoming concerns the *necessity* of the result, the fact that in normal arithmetic 2 + 2 *has* to add up to 4. That – it might be said – is what a proof should establish, and this element of mathematical compulsion lies beyond the purely empirical procedure of the naive attempt.

We will not speculate on how this criticism might be received by an advocate of the low-status method, but of one thing we may be sure: sophisticated critics of the sociology of knowledge would have no time for the empirical proof. This is because the low-status proof is just a replay of the way 2 + 2 = 4 is taught and, as we have noted, teaching must count as social determination and conditioning. Admittedly it is possible to give the process of teaching, and hence the low-status proof, a more elevated and metaphysical gloss. Perhaps when we are confronted by the demonstration with the four apples we could be said to 'apprehend the universal in the particular', i.e. see through the empirical circumstances to a timeless necessity lying behind it. Indeed, on some interpretations, this slogan might well be true. We do gain arithmetical insight by inspecting the behaviour of objects, and we do learn how to go on to handle new cases on this basis. Following our ground rules, however, we will not accept any attempt to rescue the alleged proof that gives these common-sense observations a non-naturalistic reading. The role of empirical inspection in teaching and learning is best understood in psychological and sociological terms.

Since we have not yet found an authoritative proof of 2 + 2 = 4 we have not yet fully confronted the challenge we set ourselves. Let us therefore turn to the high-status proofs of the kind taken seriously by logicians. One difficulty presents itself straight away. Most of us have never encountered such things, and yet we claim to *know* that 2 + 2 = 4. Those who insist on the significance of such proofs for our discussion would have to say that our so-called knowledge is of a second-hand and derivative kind: we can only be said to know by courtesy. We count as knowing because someone, somewhere, really knows, and they know in virtue of the proof. To defend the importance of the proof they might even have to say that we count as knowing that $2 + 2 = 4$ because someone *could*, in principle, know the proof, i.e. because $2 + 2 = 4$ is provable. Our knowledge has, so to speak, an underground connection to the proof. Without this proviso it would be difficult for them to avoid the strange conclusion that nobody really knew that $2 + 2 = 4$ until Frege and Russell had done their work in the early years of this century.[2] But whatever the real role of such proofs in the overall economy of shared belief, the fact is that they are available, they do now exist, and they call for some response from the sociologist of knowledge.

7.5 THE RIGOROUS PROOF

The proof we will consider is taken from a paper by J.L.Mackie (1966). His purpose in presenting the proof was to subject it to critical examination, and we will follow his discussion in some detail. If the proof looks difficult, this is only because the symbolism is unfamiliar. There are twelve steps in the proof, but each step is deliberately made very small in order to ensure that the proof is rigorous, and that no uncertainty creeps in. All the symbolism and background information needed to follow the argument will be given immediately after the proof itself. Here are the twelve steps:

1. $(\exists r)\ (\exists s)\ [r \varepsilon K.\ s \varepsilon K.\ r \neq s.\ (w)\ \{w \varepsilon K \supset (w = r\ v\ w = s)\}]$.
   $(\exists t)\ (\exists u)\ [t \varepsilon L.\ u \varepsilon L.\ t \neq u.\ (x)\ \{x \varepsilon L \supset (x = t\ v\ x = u)\}]$.
   $(y)\ [y \varepsilon K \supset \sim y \varepsilon L].\ (z)[z \varepsilon\ M \equiv (z \varepsilon K\ v\ z \varepsilon L)]$ – supposition.

2. $a \varepsilon K.\ b \varepsilon K.\ a \neq b.\ (w)\ [w \varepsilon K \supset (w = a\ v\ w = b)]$ – from 1, by simplification and E.I.

3. $c \varepsilon L.\ d \varepsilon L.\ c \neq d.\ (x)\ [\ x \varepsilon L \supset (x = c\ v\ x = d)]$ – from 1, by simplification and E.I.

4. $a \varepsilon M.\ b \varepsilon M.\ c \varepsilon M.\ d \varepsilon M$ – from 1, 2, and 3, using U.I. *etc.*

5. $a \neq c.\ a \neq d.\ b \neq c.\ b \neq d$ – from 1, 2, and 3, using U.I., Id., *etc.*

6. $e \varepsilon M$ – supposition.

7. $e \varepsilon K\ v\ e \varepsilon L$ – from 1 and 6.

8. $e = a\ v\ e = b\ v\ e = c\ v\ e = d$ – from 2, 3, and 7.

9. $(x)\ [x \varepsilon M \supset (x = a\ v\ x = b\ v\ x = c\ v\ x = d)]$ – from 6–8 by C.P. and U.G.

10. $a \varepsilon M.\ b \varepsilon M.\ c \varepsilon M.\ d \varepsilon M.\ a \neq b.\ a \neq c.\ a \neq d.\ b \neq c.\ b \neq d.\ c \neq d.\ (x)\ [x \varepsilon M \supset (x = a\ v\ x = b\ v\ x = c\ v\ x = d)]$ from 2, 3, 4, 5, and 9.

11. $(\exists r) (\exists s) (\exists t) (\exists u) [r \varepsilon M. s \varepsilon M. t \varepsilon M. u \varepsilon M. r \neq s. r \neq t. r \neq u. s \neq t. s \neq u.$
    $t \neq u. (x) \{x \varepsilon M \supset (x=r \vee x=s \vee x=t \vee x=u)\}] -$ from 10 by E.G.
12. $(K) (L) (M) [(1) \supset (11)] -$ from 1–11 by C.P. and U.G.

Standard text books on symbolic logic, such as Hans Reichenbach's well-known *Elements of Symbolic Logic* (1947), explain the symbolism. We will follow Reichenbach's treatment as a way of setting the stage, not only because of its clarity, but because he tried to anticipate some of the critical points that will concern us. First, though, here is how Reichenbach explains the ideas and symbols that are used in the proof.

When logicians want to say that a property f belongs to one and only one object, they write: $(\exists x) f(x) . (y) [f(y) \supset (y=x)]$

The first part $(\exists x) f(x)$ reads: 'there is at least one x such that x has the property f. The symbol $\exists$ is called an existential operator or quantifier. The dot . following it means 'and'. Later we will meet the negation symbol $\sim$, to be read 'not'. The symbol (y) means 'for all y', and is called a universal quantifier. The symbol $\supset$ is to be read as 'implies', or 'if . . . then . . .' So the whole formula reads: there is at least one object x that has the property f and, for all y, if y has the property f then that implies that $y = x$. Now let us introduce the symbol v meaning 'or'. Using this, the statement that there are exactly two distinct things, x and y, having the property f is written:

$(\exists x) (\exists y) f(x) . f(y) . (x \neq y) . (z) [f(z) \supset ((z=x) \vee (z=y))]$

The statement that the property f belongs to three and only three things becomes:

$(\exists x)(\exists y)(\exists z).f(x).f(y).f(z).(x \neq y).(y \neq z).(x \neq z).(u)[f(u) \supset ((u=x) \vee (u=y)$
$\vee (u=z))]$

And so on for 4, 5, 6 etc. distinct things.

Reichenbach then introduces the notion of a class or set of objects. Let all objects with the property f be said to constitute the set F. To say that an object x belongs to the set F we write $x \varepsilon F$. Connecting this with the formulae above we can say that if the first formula holds, then F has one member (is a 'one class' or 'one group'); if the second formula holds, F has two members (is a 'two class' or 'two group'); and if the third formula holds, F has three members (is a 'three class' or 'three group') – and so on. He then uses these formulae to define the integers 1, 2, 3 etc., which he does in the manner originally developed by Frege and Russell. The number 3, say, is defined as a property common to all classes or sets with three members (all 'three classes'). To avoid obscurity about what that common property is, we are to think of 'the number 3' as the class of all classes that have three members. Similarly, the number 2 just is the class of all classes with 2 members, and so on. Logicians deny that this definition is circular, on the grounds that being, say, three-membered (or a three-class) doesn't require the number three to appear in the formula that defines it. They argue that being a three-class is a basic property that can then be used to define the number three – and similarly, of course, with all the other integers. Thus:

$F \, \epsilon \, 1 = (\exists x) \, (x \, \epsilon \, F) \, . \, (y) \, [(y \, F) \, (x=y)]$

$F \, \epsilon \, 2 = (\exists x) \, (\exists y) \, . \, (x \, \epsilon \, F) \, . \, (y \, \epsilon \, F) \, . \, (x \neq y) \, . \, (z) \, [(z \, \epsilon \, F) \, ((z=x)$
$v \, (z=y))]$

And the same procedure can be applied to 3, 4 etc. Reichenbach calls these the 'logical definitions' of the numbers 1, 2, etc.[3]

We are now equipped to turn back to the proof and follow it step by step. Step 1 simply tells us that K and L are two-groups, with no common members, and that M is the group made by gathering the members of K and L together. The first line of the proof reads: there is an r and there is an s, and r belongs to set K and s belongs to set K, and r does not equal s and, for all w, if w belongs to K then either w equals r, or w equals s. The second line repeats this for the two elements t and u of the set L. The third line reads: for all y, if y belongs to K then it does not belong to L, and for all z, if z belongs to M that is equivalent to either z belonging to K, or z belonging to L.

Steps 2 and 3 merely repeat this information, but do so in a way that gets rid of the existential quantifiers. Instead of talking about there being (at least) one object r, reference is made to some particular object, here called a. Similarly with the other letters s, t, and u. This process of getting rid of existential quantifiers is called 'existential instantiation'. The aim is to make the content of the formulae, and the substance of their claims, more easily reorganized.

Steps 4 and 5 also get rid of the quantifiers, only here the manoeuvre concerns the universal quantifiers used in the third line of Step 1. This is called 'universal instantiation', and works on the basis that what is true for all z must be true of any particular z. Step 4 says: a belongs to M, b belongs to M, c belongs to M and d belongs to M. Step 5 says: a isn't c, a isn't d, b isn't c, b isn't d. Step 6 supposes that we are now dealing with an arbitrary member of set M called e. Step 7 concludes from Step 1 that this e must belong either to set K or set L. Step 8 then follows from Steps 2, 3 and 7, and says that e must be either a or b or c or d. Step 9 restates the reasoning of Steps 6, 7 and 8 in a form that brings out its full generality, and says that anything that belongs to M must be identical to one of a, b, c or d. Step 10 comes from combining Step 9 with Steps 2, 3, 4 and 5. It says that M is a group consisting of objects a, b, c and d, and only these.

Step 11 is got by making a move called existential generalization. In effect, this puts back the quantifiers that were taken away at the beginning. The rationale of the step is that since we have been dealing with a particular object called, say a, that belongs to M, then we can infer that there must be at least one object, call it r, that belongs to M – in symbols $(\exists r)r \, \varepsilon \, M$. This applies to the other objects s, t and u, and permits us to infer that M is a four-group. This means it belongs to the set of sets that makes up the number 4. Step 12 states in a summary form that for all sets K, L and M, if K belongs to the set of two-groups, and L belongs to the set of two-groups, and they have no members in common, and M is got by bringing the members of these two groups together, then M is a four-group. In the idiom of symbolic logic this is taken to mean that $2 + 2 = 4$.

As we run our eye over the lines of symbolic logic we seem a long way from the classroom drill of young schoolchildren learning to count and add. Here, it seems, things are thus and so because they *must* be thus and so: it is logic and logic alone that takes us by the scruff of the neck and *compels* us to take the steps of the proof and to draw the conclusion that $2 + 2 = 4$. Or is it?

7.6 THE EMPIRICAL BASIS OF THE PROOF PROCEDURE

Mackie argues that, for all the show of rigour, when we work through these twelve steps of symbolic logic we depend on exactly the same processes of thought that we encountered in the low-status proof. That is, we depend upon an instance of $2 + 2 = 4$. We use the result that $2 + 2 = 4$ in order to select, order, apprehend and arrange the very symbols of the proof. He puts it like this:

> The logical techniques used here to formulate 'K is a two-group' and 'L is a two-group' enable us to introduce 'a' and 'b' as names of the members of K, and 'c' and 'd' as names of the members of L; this ensures that the names of the members of M will be 'a', 'b', 'c' and 'd'; and the fact that there are just four names here ensures that M will be described by the expression which is a formulation of 'M is a four-group'. The proof goes through, and it yields the desired results; but it does so precisely because the theorem which we are trying to prove is true of the groups of symbols which play a vital role in the proof. (Mackie, 1966, p. 34)

The significance of Mackie's claim must not be missed. If he is right, then the logician's achievement with symbols leaves us no better off than we were when we manipulated the four apples. He has collapsed the hierarchy that would lead us to see in the high-status proof an achievement of a different order to the low-status proof. Mackie's paper is the work of a cognitive 'leveller'. Now, the sociology of knowledge itself is not in the business of levelling: its aim is to understand hierarchies, wherever they might exist, rather than to attack or defend them. But there is much to be learned from the arguments of those who do attack them: the attacks expose their conventional character. To learn from the attack we must be sure we understand it, and can assess its precise force and import.

Is Mackie advancing the claim that the logician's proof is logically defective, e.g. that it is circular? Such an impression might be gained when Mackie sums up his claim by saying that the proof 'rests upon the truth of one particular instance of the theorem being proved' (p. 35). Nevertheless, this is not his point, and he insists that his argument 'does not show that the proof is circular' (p. 35). He does not assert, for example, that the conclusion of the proof features illegitimately in the premises. Nor does he imply that a particular instance of the universal proposition $2 + 2 = 4$ finds its way into the premises. When he says that the proof 'rests upon' an instance of the theorem proved, he means that the person producing the proof, or anyone understanding the proof, needs to know how to use and recognize instances of the theorem. In short, they must know that $2 + 2 = 4$. Mackie's claim, then, does not concern the proof in isolation, but the relationship between the proof and the prover. It concerns the office or role of the proof in our knowing that $2 + 2 = 4$. This is what makes it so important for our purposes. He makes two related points: one concerns the empirical basis of the procedure adopted by the prover; the other concerns the limit this places on the contribution the proof can make to our knowledge.

As Mackie explained, the empirical and arithmetical presuppositions employed by the prover concern the introduction of the letters 'a' and 'b' and 'c' and 'd'. The twelve lines of the proof provide the setting in which we take two such letters 'a' and 'b', and then another two, 'c' and 'd'. The 'a and b' make a two-group; the 'c and d' make another two-group. The 'a, b, c, d' make a four-group. How do we know that the formulae express these specific claims about K and L? The fact is that we count them. If we couldn't count we couldn't know. We assemble 'a and b' in the rigorous proof as we assemble the first two apples in the naive proof, and we assemble 'c and d' as we assemble the second lot of apples. We bring 'a, b, c, d' together just as we brought the apples together, except that the physical act of collecting takes the form of the manipulation of symbols. Again, our knowledge that the formula that results from these manipulations is the formula of a four-group depends on our counting the a b c and d, just as we counted the apples.

The two procedures – the naive proof and the rigorous proof – thus have exactly the same inductive character. If the naive performance with apples is deemed to issue in merely empirical knowledge, then so should the same performance when it uses symbolic tokens. And if we are not going to permit the experience of manipulating apples to occasion a self-validating intuition into mathematical reality, we cannot permit it with letters of the alphabet either. It follows that the proof cannot lift our knowledge above the status of the informal, empirical procedure that we all learned as children. Without that childhood knowledge we would be powerless to construct or learn from the proof.

7.7 THE REICHENBACH DEFENCE

Logicians are well aware that their efforts might meet such a reception, and that the significance of their achievements might be threatened as a consequence. For example Reichenbach was worried that the circular-looking Russell-Frege style definition of the integers might run into trouble of this kind. He said, 'The objection has been raised against the logical definition that it makes an implicit use of empirical objects occurring in the respective number, namely in the form of signs' (p. 249). Thus the definition of the number 4 contains four existential operators. This can be seen by examining line 11 of the rigorous proof. Notice $\exists r$, $\exists s$, $\exists t$, $\exists u$ – 4 existential quantifiers: 'and it is clear that every definition of this form will contain as many existential operators as correspond to the number defined' (p. 249). This is essentially the point to which Mackie draws attention in his argument. Reichenbach, however, doesn't think it should be considered significant. If he is right he might offer a way round Mackie's observations, so it is important to attend closely to the reasons he gives.

Reichenbach tells us that although the definition of, say, the number 3 contains three existential operators it 'does not *refer* to these operators; it uses these signs but does not speak about them' (p. 244). There is no question here, he says, of the symbols being in any way self-referring:

> The correspondence between the number defined and the number of existential operators used in its definition must be regarded as accidental. It originates from a certain isomorphism between the logistic symbolism and the number classes expressed through that symbolism, and it would disappear for another form of the symbolism. (pp. 249–50)

As far as it goes Reichenbach's first observation is surely correct. He is right to say that the logical definition of the integer 3 uses the three existential quantifiers, but doesn't mention that there are three of them. The use–mention distinction shows that the definition does not fall into the error that, for example, Poincaré once complained about when he came across a definition of the number one by Burali-Forti, which itself made reference to that number (1909, p. 157). But do these considerations remove the force of Mackie's observations? No, they leave them untouched. While it is true that the definitions of the integers do not reflexively mention the number of quantifiers in the definitional formula, this is not true of the thought processes required of the prover. Mackie's point is that the proof user employs these arithmetical facts about the structural features of the symbolism when, all along, that symbolism is meant to be furnishing the prover with justification for such an employment. The correspondence between the number defined and the number of existential quantifiers in the definition is therefore *not* accidental but essential to the way the prover puts such definitions and proofs to work.

This bears upon the question of circularity. The proof itself may not be circular, but there surely *is* a circularity involved in its use. If someone is seeking to justify a principle, and in the course of their justificatory argument has to have resort to that very principle, then they can be called to account over the soundness of their procedure. If the principle is deemed to lack justification until the argument is finished, then its employment in the conduct of the argument is thus far unjustified. For the charge of circularity to stick it does not matter in what precise capacity the principle operates, whether as a premise or in some more tacit role. All that matters is that the process of argument genuinely depends upon it, and cannot go forward without it. It is in this spirit that those who would justify induction must avoid *all* dependence on inductive principles, and those who would justify deductive principles such as *modus ponens* $(p.(p \supset q) \supset q)$ must likewise ensure that they don't secretly utilize any such reasoning at any place in their justificatory procedure (cf. Haack, 1976).

The conclusion is that once we consider both prover and proof there is a circularity to be detected. The prover utilizes an instance of the theorem to be proven. How is appeal to that instance to be justified? Short of making implausible claims about direct insight into these particular instances where $2 + 2 = 4$, the justificatory challenge can be met in no other way than by appeal to the general theorem that $2 + 2 = 4$, which was the very thing to be proven. The justificatory use of the proof is thus compromised, and it was in that capacity that critics of the sociology of knowledge appeal to the proof.[4]

From a naturalistic and explanatory perspective (as distinct from a justificatory one) this circularity is not something to be deplored or avoided, but something to be described and understood. It is a phenomenon of interest in its own right. But if the proof we have examined does not generate and establish our knowledge that $2 + 2 = 4$, then how *is* it generated and established? If we construct the proof that $2 + 2 = 4$ on the basis of our childhood knowledge that $2 + 2 = 4$, what is the basis of that? We shall now turn briefly to the naturalistic and sociological answer to that question.

7.8 THE SOCIAL FOUNDATIONS OF ARITHMETIC

We routinely utilize the equation $2 + 2 = 4$ because it is part of a conventionally accepted practice. It is grounded in a shared way of ordering and sorting objects, where these ways have become fixed and paradigmatic patterns in our collectively sanctioned thinking. To see what this entails we need only go back to the apples of the naive proof. We are familiar with some of the behaviour of these objects through our everyday experience of them, but this alone doesn't yield arithmetic. The empirical facts that are relevant to arithmetic are facts about the behaviour of such objects under rather special conditions, namely, when subject to certain characteristic patterns of manipulation (think of learning to count and to add, take away and divide by using the beads on an abacus). These techniques have to be rigidly adhered to, and this is where we need a collectively agreed system of conventions.

Wittgenstein made the point elegantly for the case of $2 + 2 = 4$ when he imagined someone who says, 'You need only look at the figure

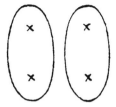

to see that 2 + 2 are 4'. In that case, replied Wittgenstein, you need only look at the figure

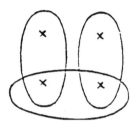

to see that 2 + 2 + 2 are 4 (Wittgenstein, 1978, I, 38). His point is that when we look at the four crosses – or any other four objects – our act of looking must be an educated one. It is false that we 'need only look'. We need to analyse the figure and have at hand a very specific way of relating it to the symbols 2 + 2, otherwise we will not extract from it what, to us, counts as its mathematical significance. To extract the mathematical essence of the figure requires that the technique we use has the character of a convention. It must be a shared technique that can be taken for granted in our dealings with others, and whose proper use is collectively agreed and sanctioned. It was this convention that was already in place and was put to use in the construction of the rigorous proof.

Such an account can offer a ready explanation for the quality of 'necessity' or 'compulsion' that attaches to arithmetical operations. Mathematical necessity is just a species of the moral necessity that frequently attaches to the more important social conventions. Thus the inexorability of mathematics is just the inexorability of the social demand that we deploy our techniques like this rather than like that, e.g. so that 2 + 2 = 4 and not 2 + 2 + 2 = 4. If we wish to find an illuminating comparison to help us understand the use of words like 'must' and 'can't' in connection with mathematical operations, we should examine their use in the context of other conventional activities like following the rules of a game. Talk of 'must' and 'can't' is not mysterious, it is just the language used to *induce* conformity in the individual player.

Are we saying that if we had so desired we could have had an arithmetic in which 2 + 2 does not equal 4 but equals, say, 5? To some critics this seems so outrageous, and so redolent of Orwell's *Nineteen Eighty-Four*, that it shows the absurdity of any approach which gives convention a significant role in arithmetic. The reply is that this allegedly outrageous consequence does indeed follow from the approach adopted here, but it is not in fact outrageous at all – and nor does it involve brainwashing or the violent disruption of the normal working of our brains. The specific challenge of 2 + 2 = 5 has been answered briefly but effectively in Lakatos' *Proofs and Refutations* (1976). In the course of an imaginary dialogue between two mathematics students called Gamma and Kappa, we find that Gamma is attracted to the programme of seeking clearer and clearer definitions of the terms used in a proof. In the interests of rigour Gamma wants to make them so clear that there can be no dispute over their interpretation. To explain what he means he adopts the same line as our critic, and appeals to 2 + 2 = 4 as a paradigm case of an indubitable, secure, stable mathematical truth. Gamma is made to declare: 'There is nothing elastic about the meaning of these terms, and nothing refutable about the truth of this presupposition' (p. 101). The reply by Kappa (who has been reading the French mathematician Felix, 1960, pp. 2–3) is swift:

> *Kappa* But this is child's play! In certain cases two and two make five. Suppose we ask for the delivery of two articles each weighing two pounds; they are delivered in a box weighing one pound, then in this package two pounds and two pounds will make five pounds!
> *Gamma* But you get five pounds by adding *three* weights, 2 and 2 and 1!
> *Kappa* True, but our operation '2 and 2 make 5' is not addition in the originally intended sense. But we can make the result hold true by a simple stretching of the meaning of addition. Naive addition is a very special case of packing where the weight of the covering material is zero. We have to build this lemma into the conjecture as a condition: our improved conjecture will be '2 + 2 = 4 for "weightless" addition'. The whole story of algebra is a series of such concept- and proof-stretchings. (pp. 101–2)

If this supports the view that we *could* have a convention in which 2 + 2 = 5, then it will be asked: why *don't* we have it? Why do we say 2 + 2 = 4 rather than 2 + 2 = 5, or all the other things we apparently could say? The objection lurking behind this query is that our current mathematical conventions might be more than 'just' conventions. They might have been selected or reinforced because they 'correspond' to some truth, or because they are informed by some uniquely rational virtue that singles them out.

The naturalistic response must be to take the question seriously, but to insist that if it has an answer it will not be in terms of our practices 'corresponding' to some mysterious mathematical reality. It will be for some naturalistically explicable psychological and social reasons. Consider, for example, why we might have a preference for what Lakatos called 'weightless' addition (where 2 + 2 = 4), rather than one of the indefinitely large number of alternatives. A sociological answer might appeal to principles of the following kind: to establish a convention for addition means solving a coordination problem, that is, it means getting everybody to adopt the same procedure. Coordination problems are easier to solve if they have a 'salient solution', one that is automatically visible to everyone and where everyone routinely assumes that it is visible to everyone else. Salient solutions are often extreme solutions, ones which lie prominently at the beginning or end of the spectrum of alternatives. Weightless addition may be such an extreme and prominent solution. There are therefore pragmatic reasons connected with the organization of collective action that would favour saying 2 + 2 = 4, rather than 2 + 2 = 5 or 6 or 7 or . . . . As a convention it is probably easier to organize than the others, and therefore more likely to arise historically.

This argument should be seen as an illustration rather than as an historical conjecture. Answering questions of the form: why do we have convention X rather than Y? is notoriously difficult and, in practice, frequently impossible. The point, however, is that such historical processes are in principle intelligible without stepping outside the bounds of naturalism. Within those bounds the sociology of knowledge can make contributions of the kind just sketched.

We set out to answer the question: how can a sociological account of 2 + 2 = 4 be called for, given that it can be proven? The answer is that the proof requires the prover to know that 2 + 2 = 4, so the real basis of that knowledge still awaits elucidation, and the outline account we have offered is sociological. There is, however, a further point to notice. The proof ends with a statement that has the form of an *implication*: roughly 'if there is a two-group, and another two-group, then there is a four-group' or '(∃2).(∃2) ⊃ (∃4)'. But the proof was meant to prove an *equation*, namely, 2 + 2 = 4. Implications and equations are different things. They may both have two sides, but they don't work in the same way. Implications can be transformed into strings of conjunctions and negations: equations can't. Equations can be added, subtracted and multiplied; implications can't. So as well as taking for granted what it is said to prove, it isn't even clear that the proof gets us to the goal we were aiming at. It is open to argument whether the proof really proves what it is usually taken to have proved.

Friedrich Waismann uses these grounds to deny roundly that proofs of the kind under consideration can possibly prove an arithmetical formula. They may establish that a certain implication is a logical truth, but not that the equations of arithmetic are true. In fact Waismann anticipated the conclusion that we have drawn from Mackie: 'The logical truth', he said, 'is not used, and cannot be used, to prove the equation, but the equation (or: the rule of addition) is used to prove the logical truth' (Waismann, 1982, p. 65).

The position is an intriguing one. How do we decide when two statements, admittedly of a somewhat different form, 'really' assert the same thing or a different thing? Does the proof issue forth the arithmetical truth that we wanted to see proved or doesn't it? Does the last line really mean 2 + 2 = 4, or does it mean something else? The clear fact of the matter is that different authorities make different judgements: some see an essential identity where others see a fundamental difference. From the standpoint of the sociologist of knowledge the question to ask is not 'who is right?' – both sides have their reasons – but: 'what sustains, or could sustain, the divergent judgements?' Who adopts one strategy, and who adopts the other? We have now uncovered a further sociological dimension to the proof that 2 + 2 = 4. Waismann's alternative reading of the last line has revealed for us the conventional character of the *interpretation* of the proof. It appears that the meaning attributed to its conclusion, the subject matter to which it is deemed relevant has, when stable, the character of an institution. For example, if the proof has a foundational role, that is, if it is seen as providing the materials out of which arithmetic is built, and the basis on which it can be justified, then this too is a matter of how it is treated by its users. It is a status given to it for which the users of the proof must be held responsible, not the proof itself.

Not all proofs are like the rigorous proof of 2 + 2 = 4. Not all proofs add so little to our knowledge. In order that our chosen example should not give too negative an impression about the role of proofs in general, it is appropriate to indicate the more positive role that proofs can play.

The typical proof contributes new techniques to the area under study, it is genuinely informative and enabling in a way that our example was not. The new techniques will often be analogies imported from another area of mathematics or, at a more basic level, novel empirical operations that serve as new patterns of inference and thought. This explains why proofs can teach us new things, can surprise us, and why they are valued as genuine contributions. Historical case studies provide the best approach to these topics, bringing out both the empirical and conventional sides of the story. For example, many of the techniques of the differential and integral calculus came from thinking about how shapes might be built up out of very thin slices ('infinitesimals') that must be added together (Boyer, 1959). Likewise, a proof of Euler's theorem about the relation between the vertices (V), edges (E) and faces (F) of a polyhedron (V − E + F = 2) can come from imagining the polyhedron made of material that allows a face to be removed and the remainder stretched to make a flat surface (Lakatos, 1976). Again, to revert to the example mentioned at the beginning of the chapter, the statistical concept of correlation was originally structured around a model of biological inheritance. Different evaluations of the importance of heredity also lay behind the disputes about how the new mathematical concepts were to be extended to new areas of study (MacKenzie, 1981).

In order to convey in a simplified and accessible form the lesson that emerges from such historical studies, we can make use of an elementary but hypothetical example. The example again comes from Wittgenstein (1978), and shows how proofs grow out of empirical procedures. Consider the proposition that it is possible to construct a rectangle from two parallelograms and two triangles. This, for all its simplicity, has some of the hallmarks of a typical mathematical theorem. Here, said Wittgenstein, is the proof (1978, I, 50):

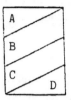

A and D are two triangles; B and C are two parallelograms. The idea of the proof is grounded in empirical operations such as cutting or folding paper, sawing pieces of wood or drawing lines with a pencil and ruler. This constitutes the empirical core of the proof procedure. The proof also furnishes us with a technique that we can carry over to new problems. For instance, says Wittgenstein, we can now easily see how to construct a long straight strip with parallel edges from a large number of parallelograms. This might be called a new theorem. Here is the proof:

We could now apply this theorem and make a prediction about the effect of conjoining a large number of pieces of wood or metal shaped as identical parallelograms. They should form a long, straight parallel-sided strip. The prediction is an empirical matter, and might well fail. When the experiment is tried and the strip built, its edges might be curved rather than straight. Here we have a choice. On the one hand we can treat the whole process as empirical, and say that these pictures are misleading and inaccurate. On the other hand we can say that the pictures are untarnished by the failure. We can decide to keep the pictures as sources of insight. We can, as it were, put them in the archive. We would then conclude that some systematic error must have disturbed the conditions of the experiment: the manufacture of the parallelograms cannot have been accurate. They must have been slightly out of alignment. The 'must' and 'cannot' betray what is happening, and should be enough to alert the sociologist of knowledge. The pictures are being turned into a standard or norm. They are being given a particular use. Their function is no longer to describe reality, but to act as a measure of it. If their employment in that role becomes a convention, the empirical model will be accorded the status of a proof of what has to happen – that is, happen with mathematical necessity. Critics of the sociology of knowledge think that mathematical necessity removes a proposition from the realm of the social; in fact, it signals its transition into that realm.

7.10 SELF-EVIDENCE

It may be useful to conclude with an inventory of the different ways in which the sociologist of knowledge may, as it were, gain entry into the processes of mathematical proof. To some extent these possibilities have been illustrated and realized in the discussion in the proof that 2 + 2 = 4, but we have seen that, in certain respects, that proof is not wholly typical. It is therefore appropriate to address the problem in more general terms. This can be done by reflecting on the formal and structural properties of proofs. A proof is an argument that is meant to establish that its conclusion is true. Like any argument it must have premises. The proof then demonstrates in a rigorous way that the premises imply the theorem in question. It is for this reason that a proof can be divided into two parts:

i) premises
ii) deductions

In many cases the premises of a given proof will themselves be the proven consequences of a previous argument, but on pain of infinite regress, the process must have a beginning. At the beginning there will have to be premises that are themselves unproven. If the theorem that eventually follows from these ultimate (unproven) premises is to be considered genuinely known and fully established, then the premises themselves must be known to be true. They too must be fully established, but by virtue of some process other than proof. They cannot be mere assumptions, because otherwise that conjectural status would be transmitted down from the premise to the conclusion. The theorem itself would have a merely conjectural status rather than being, in any conclusive way, really known. For this reason the basic premises of a mathematical system (e.g. the ultimate starting points of Euclid's geometry) were traditionally taken to be self-evident. For example, is it not self-evident, that given a straight line AB, and a point P, not on the line, only one straight line can be drawn through P that is parallel to AB? It was required that premises such as this possess a transparency and simplicity that would allow any rational person to see or intuit their truth a priori.

Here, immediately, we see a point of entry for the sociologist. So called 'self-evidence' is historically variable. Propositions that are counted as self-evident at one place and time are not always accorded that status by other people at other times and places. Thus historians of mathematics have recorded the long-standing but fluctuating unease about the self-evident premises of Euclid's geometry. But historical variability, by itself, does not immediately establish the relevance and presence of social variables. For that conclusion to follow it is necessary to adopt a particular perspective on the variability or, to state the point in another way, it is necessary to remove certain prejudices. It is always possible to gloss the variable attitudes towards allegedly self-evident propositions as a fluctuation between truth and error. Sometimes, it can be said, their truth is clearly apprehended, while at other times it slips from our grasp and our vision is clouded by error. This reduces all historical variation of the kind in question to a mere tug of war between (alleged) truth and (alleged) error. The standpoint that has informed this book is somewhat different, and suggests a different perspective on historical variation. It should be viewed in a detached and naturalistic fashion. Rather than endorsing one of the claims to self-evidence and rejecting the other, the historian can take seriously the unprovability of the claims that are made at this level, and search out the immediate *causes* of the credibility that is attached or withheld from them. Self-evidence should be treated as an 'actors' category', and as a phenomenon whose imputation must be explained in terms that are quite different from the 'insights' and 'intuitions' that are often invoked and spoken about by the actors themselves.

In the case of the axiom of parallels mentioned above, historians have not only described the difference of opinion on its status as a self-evident intuition about the properties of space: they have on occasion been able to go further and identify certain interests that seemed to have informed these different perceptions. Thus Richards has shown how the nineteenth-century empiricist advocates of non-Euclidean geometry, who insisted that the axiom of parallels might be empirically false rather than a priori true, were using the doctrine as part of an ideological strategy. This indeed was the source of the attraction of non-Euclidean geometry to them. It was a means of undermining the claims of institutions such as the Church which stood in the way of the newly emerging profession of science. As long as theologians could point to geometry as a source of a priori knowledge about reality, it proved that such knowledge was possible, and hence supported their claim to be able to supply further instances of this commodity. It meant that Euclidean geometry lent indirect support to older and rival sources of cognitive authority that were felt to oppose the growth of science. Non-Euclidean geometry thus became a vehicle for an aggressive professionalizing ideology of naturalism (Richards, 1979, 1988; for background, see Turner, 1974).

Given the generality of the structural fact that every argument needs premises – with its implication that every proof must be grounded in what cannot be proven – we would not expect such historical investigations, of the kind described above, to be confined to any one branch of mathematics, such as geometry. We might expect them to be applicable to, say, set theory as well. One of the triumphs of contemporary pure mathematics has been the success of the programme of reducing mathematics to set theory. Branch after branch of mathematics has been shown to follow from the theory of sets, and this theory has itself been reduced to a small number of axioms, though there is more than one way of achieving this axiomatic organization. The process of reduction has meant that various disputes and problems in mathematics can be reformulated and relocated as problems in the underlying theory of sets. They have been turned into disputes about how set theory should be axiomatized, and about what should or should not be accepted as an axiom.

Consider, for example, the following quotation from the eminent French mathematician Emile Borel (1871–1956). In 1905 he asserted that,

Any argument where one supposes an *arbitrary choice* to be made a non-denumerably infinite number of times . . . [is] outside the domain of mathematics. (Quoted in Moore, 1982, p. 85)

Borel, along with many other mathematicians of the time, was objecting to the procedure that became enshrined in the so called 'axiom of choice', formulated by Ernst Zermelo (1871–1953). This axiom is at the basis of modern set theory. The claim made by the axiom can be illustrated very simply. Suppose five children A, B, C, D and E each have a set of toys. It is clearly possible to create a new and distinct set or grouping of toys by arbitrarily selecting one item that belongs to

A, and one that belongs to B, and so on, until we reach E. No one has any problems with this easily imaginable and common-sense procedure. However, in many branches of mathematics, it is both tempting and useful to imagine such processes applied not to five sets, or even to a large, finite number of sets, but to an *infinite* number of such sets. The sequential process of choice is even held to make sense when the number of sets to be chosen from is so great that there is one for each real number. This is the non-denumerably infinite number of arbitrary choices to which Borel referred. So the question is: can we take for granted an analogy between the finite and the infinite case, and proceed in this way, or talk of proceeding in this way, with a good conscience? Borel was returning a negative answer to this question, and was doing so in the context of a protracted debate with Zermelo. In this debate Borel was initially in the majority, but Zermelo eventually won in the sense that the sheer utility of his axiom eventually secured it a place within acceptable mathematical practice. References to supposed infinite sequences of arbitrary choices became explicitly acknowledged and ever more widely used in the construction of acceptable proofs. The efforts of Borel and others to keep such processes 'outside the domain of mathematics' failed.

This example, which has been described in fine historical detail by Moore (1982), not only serves to illustrate the variability of mathematical intuitions of self-evidence but also invokes the theme of boundary maintenance introduced in Chapter 6. It dealt with the disputed issue of what belongs 'inside' a science and what should be kept, or cast, 'outside'. How do mathematicians decide what is mathematical and what is not? The problem is endemic. There are a number of different ways in which the boundary can be drawn depending on which aspects of mathematical practice and which mathematical methods are deemed central, or definitive, or paradigmatic. For instance, mathematicians have long been aware of the difference between lines of reasoning which show that some result must exist, and others which go on to show precisely what that result is. Thus, if we are told that a continuous line drawn on a piece of graph paper has a negative y value at $x = 1$ and a positive y value at $x = 2$, we know that it will need to cross the x axis at some point, but we do not know at which point. Some mathematical proofs of existence are like this, and opinion has long been divided on the relative merits of such 'pure existence proofs' by comparison with 'constructive' proofs, i.e. proofs that exhibit the result referred to in the proof. Deciding that pure existence proofs are suspect, and attempting to push them outside the domain of fully acceptable mathematics would be one way of taking a stand on a matter of self-evidence. Some mathematicians declare that it is self-evident that constructive proofs are superior and, indeed, that they alone are truly acceptable. Others find it equally self-evident that pure existence proofs are acceptable. Such commitments are not themselves susceptible to proof because they concern what is to count as an acceptable basis for offering and accepting proofs. Many classical mathematical results depend on pure, non-constructive lines of reasoning, and it would create something of a crisis in mathematics to outlaw them. But then, perhaps a crisis is just what is needed if self-evidently unacceptable standards of reasoning have been permitted to enter into mathematics. Such results as deserve to be salvaged can be given new, constructive proofs. The rest are perhaps best discarded. This was the general character of the crisis that struck the foundations of mathematics in the years after the First World War. The origins of the crisis can be traced back at least as far as the arguments over Zermelo's axiom, but the fact that it reached such a pitch of intensity shortly after 1918 may be no accident. Forman (1971) has documented what almost amounted to a desire or hunger on the part of many in the mathematical and scientific culture of central Europe in those years to participate in a cultural crisis. The intriguing possibility is that once there is a motive to produce a crisis, there are plenty of ways in which this can be done, even in a field such as mathematics. The fact that mathematical results rest on proofs gives them no special protection in this regard. All that is needed to produce a crisis is a shift in the threshold of what counts as self-evidence.

A further example from set theory will reinforce these themes. Consider the question: how many points are there in a straight line? It is easy and natural to speak of there being specific points 'on' a line, e.g. the horizontal line or axis on a graph paper. Thus the line runs from $x = 0$ to $x = 1$, and there can be said to be a point at, say, $x = 0.5$. It seems a small step from accepting that there are points on a line, to thinking of the line itself being 'made up' of points. But if it is made up of points, won't there be a specific number of points that constitute the line? Mathematicians such as Georg Cantor (1845–1918) wanted to answer this question in the affirmative and developed techniques which permitted a determinate answer to be given – or, to proceed more cautiously, they developed techniques which can be interpreted in this light, and they took it as evident that they should be so interpreted. Specifically, Cantor introduced the infinite cardinal number $\aleph_0$ (read as *aleph nought*) which is the number of the totality of natural numbers (i.e. 0, 1, 2, 3 . . .). He conjectured that the number of points on a line is $2^{\aleph_0}$. This is itself an infinite cardinal number, but a different one from $\aleph_0$. Cantor called it $\aleph_1$, and conjectured that it was the 'next' infinite cardinal after $\aleph_0$. Cantor's so called continuum hypothesis, that $2^{\aleph_0} = \aleph_1$, has yet to be proven true or shown to be false. Cantor himself treated this hypothesis as self-evidently true; but at least one leading expert in the field has recently expressed the view that mathematicians may come to view it as 'obviously' false. The expert in question is P. Cohen. For a clear discussion of Cantor which can serve as a valuable introduction to more detailed studies such as Moore's, see Tiles (1989 and 1991).

These disputes bring out clearly the need for what earlier in the book we have called a finitist account of meaning. They involved a continuous renegotiation of ·the terms: set, class, real number, infinity, count, function, definition, construction, continuum, etc. They also reveal some intriguing parallels between disputes in mathematics and in the physical sciences. Any given set of empirical observations and experimental procedures must be interpreted. Meaning must be attached to them and there is no unique way of reading their significance. The same applies to symbolic manipulations. They too require an interpretive framework, and in the context of an alternative framework may have a quite different meaning. If Cantor was indeed telling us how many points there are on a line, it was in virtue of his attaching a new meaning to the question. Questions and answers alike exist in an interpretive context, and that context requires an agreed pattern of practice amongst the relevant group of professionals. As with the analogous case of geometrical axioms, disputes over set theoretical axioms provide much suggestive raw material for sociological analysis. Fine-grained historical description has yet to be followed up by a full sociological analysis. Having been admirably documented as an intellectual phenomenon, it still awaits further analysis as a cultural phenomenon.

So far the discussion has centred on the first of the structural properties of proofs – their self-evident, unproven premises. What can be said about the other structural component, the deduction? Some critics of the sociology of knowledge, realizing that the starting points of an argument may be historically and hence socially variable, place all of the weight on the deductive links in a proof. They say, in effect, 'Forget the premises, the real substance of a proof lies in its logical form. Proofs say: *If* such and such is assumed, *then* such and such follows. If you grant this, then you *must* concede that this further thing follows'.

On this view, the sociologist is allowed a role in explaining the choice of premises, but is said to have nothing to offer on the topic of deduction itself. This position represents a retreat from the more traditional view because it gives up the necessary injection of self-evident intuition into the premises of the proof, and hence compromises the status of the conclusion of the mathematical argument. Nevertheless, it is a widely held view of proof, and is often employed against the sociologist of knowledge.

Two things can be said in response. The first is to point out that the deductive steps in mathematical reasoning have to be properly organized. A proof isn't any set of deductions: it is a set of deductive links that has coherence and direction. That coherence and direction come from the reasoner having some underlying plan. There must be an idea behind the proof. It was in order to illustrate this that we drew attention in section 7.9 to the empirical analogies and models that have played a significant role in the history of mathematics, e.g. the model of infinitesimal segments, or the model of stretching a shape flat in the proof of Euler's theorem. The analogies that guided Cantor and Zermelo between finite procedures of counting and finite sequences of choices and their infinite counterparts, represent further instances of the use of such empirical models. The example from Wittgenstein of the proof that a rectangle can be made from two triangles and two parallelograms provided a further and extremely elementary illustration of such a proof idea. The proof idea can provide the basis for the intuitive understanding of a proof – that is, the flash of insight that can run ahead of the step-by-step articulation of the deductive reasoning. That is how it guides and organizes the individual steps. The striking fact about the rigorous proof of $2 + 2 = 4$ was that the proof idea – the underlying model governing the line-by-line deductive process – was itself the empirical knowledge that $2 + 2 = 4$. That was why the proof added nothing significant to our understanding. There was here no model transferred into a new area of application.

Naturally the psychologist, as well as the working mathematician, will have a professional interest in the general operation of proof ideas. They can help shed light on the mysterious processes of mathematical creativity: see Polya (1954) and Wertheimer (1961). But as well as their psychological significance proof ideas also have a sociological interest. A proof idea, particularly one that has received detailed deductive elaboration, provides a new resource. One successful application can provide the model for another, slightly modified employment. Traditions can be built up, and the very success of such a tradition can lead to the marginalization or even exclusion of alternative proof ideas. A striking example of such exclusion is the preference for geometrical approaches (rather than more formal and analytical approaches) to the calculus in eighteenth- and early nineteenth-century British mathematics. This was a battle that was fought and won by the Cambridge formalists, such as Peacock and Babbage, who championed the new analytical approach; and it was a battle that had to be fought and won all over again in the nineteenth-century Scottish universities (Enros, 1981; Bloor, 1981; and Davies, 1961).

Identifying the role of the proof idea is one way in which processes of a sociological significance can be located within the texture of the deductive reasoning of a proof. Historically, proof ideas can be thought of as operating in a similar way to Kuhn's paradigms (Kuhn, 1970). But even so suggestive a parallel as this may not satisfy a critic fighting a determined rearguard action against the sociology of knowledge. Such a critic would point out that no matter what might be discovered about the overall organization of the deductive steps of a proof, the steps themselves must have a special property: they must be *valid*. Validity, the critic would insist, is a purely logical property, not a sociological property. How, then, is validity to be defined? A valid form of argument is one that never takes a reasoner from true premises to false conclusions. If the premises of a valid argument are true, then the conclusions must be true. Proofs, it might be said, are essentially just valid argument forms. For example, if it can be shown that the reasoning of a proof has the form called *modus ponens* then it would appear to be valid. According to this principle, if p implies q, and you grant p, then you must grant q: symbolically $(p \,.\, (p \supset q) \supset q)$. It is at the level of the ultimate deductive steps themselves – steps such as *modus ponens* – that the last battle must be fought. It is here that the rationalist philosophers, opposing the advance of the sociologists of knowledge, defend the last ditch. Sociologists must therefore reply by demonstrating how their perspective can gain a purchase on the most elementary and basic of all deductive steps.

Most of us take *modus ponens* to be valid, i.e. as something that will never lead us from truth into error, because it appears self-evident. But how do we know it will never let us down? Can we prove that it won't? If we try to prove it, what principle of reasoning shall we use in this proof? Suppose we were to look for some sign or indicator of validity, call it v; and suppose we were to locate some plausible candidate (e.g. that the symbols expressing *modus ponens* constitute a truth table tautology).[5] Could we reason as follows? If an argument form has property v, then it is valid: and *modus ponens* has property v, therefore it is valid. Clearly not, because this argument itself uses *modus ponens*. We cannot prove *modus ponens* valid by using *modus ponens*, for that would presuppose the very thing we were trying to prove. It would be to argue in a circle. We might of course ask: why shouldn't we argue in a circle? The answer is because, if we are allowed to justify what we take to be valid inferences by using these self-same inferences, we could not object to someone using equally self-evident *invalid* inference forms to justify *invalid* inferences. Our circular 'justificatory' procedure would have gained us no advantage, and no basis for privileging our preferred inference form (Haack, 1976).

Although the discussion above has been developed around the special case of *modus ponens*, the argument is completely general. Deduction cannot be justified in any absolute way. This does not mean that we should eschew it, but it does mean that its sources of credibility must be re-examined with an open mind. It brings the whole subject of logical compulsion back into the sphere of our naturalistic curiosity. It also brings our discussion of proof full circle. Instead of the need to study two separate aspects of proof (the self-evident premises and the subsequent deductions) there is really only one problem: namely self-evidence. In particular: why do we find elementary deductive steps so compelling and self-evident? Part of the answer may well be biological. Perhaps we have evolved in such a way that we naturally act according to *modus ponens*. Presumably this form of inference has proved its general reliability and utility as a way of combining information. To some theorists, this may suggest that we have evolved an innate 'language of thought' with the appropriate logical rules of symbol manipulation, such as *modus ponens*, 'hard-wired' into our heads. This is, however, only one possibility among others, and nobody yet knows the ways in which our logical intuitions are internally represented. But whatever their internal representations, it is likely that certain basic steps of reasoning, in some guise or other, are innately attractive to us as part of our natural rationality (Barnes, 1976).

The sociologist's interests in these problems will focus not on their evolution or psychological representation, but on the conventionalized structures that are built up, collectively, out of these innate individual tendencies (Barnes and Bloor, 1982). For example, how are the different reasoning tendencies selected for use in any given case? How are they combined into systems? How and where are they deemed to be rightly brought into play? What is their proper scope? These might loosely be referred to as the *normative* aspects of the phenomenon. They concern what Durkheim would have called the 'collective representation' of our reasoning processes, rather than their 'individual representation' (Durkheim, 1898). Whatever its basis, the fact is that our natural disposition to follow *modus ponens* can sometimes carry us from true premises to false conclusions. Followed in a simple, mechanical and faithful way, it can generate manifestly unacceptable consequences. If we stare intently at the formula 'p, if p then q, then q', it can appear that it *must* symbolize a flawless and totally reliable form of inference. How could such steps ever lead us astray? And yet, it has been known since antiquity that they can do just that.

The classic formulation of the difficulty is known as the paradox of the heap. If you begin with a heap of sand, and remove one grain, you still have a heap. So, remove a grain – you still have a heap. Now remove another grain, and so on. If the premise is true, you will still have a heap; and, of course, at the beginning of the process, you still do. Removing two grains does not turn the heap into a non-heap, but repeating the process time after time clearly will. The logical significance of this well-known story is that the argument can easily be cast into the form of a repeated series of steps, each one of which has the form of a *modus ponens*, with true premises. The premise that, if you have a heap and remove a grain, you still have a heap, provides the conditional part of the *modus ponens*: if p then q. Removing a grain adds the premise p; and the conclusion q is the proposition that you still have a heap. In other words, repeated application of modus ponens can take us from true premises to false conclusions, because each step leaves us with the conclusion that we still have a heap of sand, no matter how many times we have removed a grain. So, according to our definition of deductive validity, *modus ponens* is not valid after all (Sainsbury, 1988).

Modus ponens is among the most simple and central of all the principles of deductive inference. If this cannot be rendered absolute, then surely nothing can be. We can therefore see that its apparent self-evidence must have been an artefact, the product of our limited vision and imagination. To the sociologist this naturally suggests that its status is that of a local convention. Logicians nevertheless typically insist on taking *modus ponens* as valid rather than as invalid, as they would have to do if they had taken the counter-example of the heap seriously. Of course, all logicians know about the problem of the heap, but they usually insist on classifying it as a 'paradox' rather than as a counter-example. Where it is not ignored entirely it is put into a special category, and counted as an oddity or as itself embodying some kind of fallacy. (See for example its treatment in the classic textbook of Keynes, 1906, pt III, ch. VII, §324.) *Ad hoc* reasons are found for laying the blame on the details of the example, e.g. on the vagueness of the concept of a heap, rather than on the principle of reasoning itself. This is, of course, very understandable. The principle of *modus ponens* is so useful that no one could totally dispense with it in practice. The problem is where and when it is proper to use it, and how to draw a boundary around its 'correct' uses, to protect it from association with incorrect uses. The problem is like that of restricting the application of the commandment 'Thou shalt not kill', to ensure that it does not yield the 'wrong' conclusion, e.g. the inconvenient implication that wars are wrong. A sociologist notes that these boundaries of proper application have the status of conventions, because they are not pre-programmed into our brains, nor are they ultimately justifiable by reference to any absolute principles. Nevertheless, they are typically felt to be 'obvious' or 'self-evident'; so self-evidence becomes a sociological category. It can be seen as a reference to the taken-for-granted features of a local cultural tradition. As such, there is a call for a sociological explanation just as much in the cases of rigorous logic and sophisticated mathematics as when we are dealing with standards of behaviour or deportment or dress, or any other clearly acknowledged social and cultural phenomenon.

Conclusion

At the start of this book we said that our intention was to produce a basic introduction to the sociology of scientific knowledge, concentrating on fundamental themes and avoiding evaluation of the many and diverse recent developments which currently build upon these themes. Having reached the end, the reader will now be able to judge how far this objective has been achieved. Perhaps he or she will feel suitably equipped by the grounding this book provides to venture into the many debates which currently engross the field. Or, alternatively, perhaps he or she will be wondering whether immediate involvement in these debates would not have represented a better use of time, without the distraction of the preamble provided here. Naturally, we ourselves believe strongly in the value of a thorough and richly detailed working through of the issues considered here for anyone who intends to take a serious interest in the sociology of science. It can be counterproductive to go ill equipped straight to the current research front with its numerous differentiated perspectives and endless controversies.

We also said at the start that the basic picture to be presented would neither be the dominant position in the field as a whole nor a review of the variety of pictures therein: it would represent our own point of view. As it happens, there are reasons for believinhg that the general features of our own position are particularly well suited to the aim of the book as a basic introduction to the field. Our position has long been a naturalistic or materialistic one. When first developed two decades ago it proceeded simply by putting the same queries to the beliefs and concepts of the sciences as were already put by sociologists and anthropologists to other beliefs and concepts (Barnes, 1974). The effectiveness of this derived from the fact that the prevalent rationalism and dualism of the time was insisting on special treatment for science and on a taboo against its being faced with such queries. There was thus nothing epistemologically radical about the approach being taken: it was not necessary to question whether there was any such thing as 'reality' or 'experience' in order to proceed. Radical conclusions could be derived from epistemologically routine positions simply by applying them with (atypical) consistency. Thus our position always took for granted that, as it were, 'experience' and/or reality *were there*, that 'no consistent sociology could ever present knowledge as a fantasy unconnected with men's [*sic!*] experiences of the natural world' (Bloor, 1976, p. 29; 2nd edn 1991, p. 33). The value of this approach in the present context is that it keeps sociology in touch with other fields that study human beliefs and dispositions – psychology and biology for example – and reminds us that sociology as currently constituted can only seek to offer a *partial* view of what creates and sustains knowledge and the distribution of its credibility.

More recent developments in the sociology of scientific knowledge have almost all been idealist rather than materialist in their orientation. Why this has been so it is difficult to say. Perhaps it has been helpful to the narrowing of vision which academic specialities always seem to engage in as they develop: an idealist approach, which denies that speech has referential functions and that there is anything (other than itself) which knowledge is *about*, usefully equates all that there is in the world with what the sociology of knowledge studies, with no problematic residue or remainder, nothing lying beyond the sociological gaze. Alternatively, perhaps the move to idealism should be seen as part of a general trend running through all of the social and human sciences and more widely through our culture as a whole. And indeed just such a general trend does seem to have occurred over the last two decades and it is a fascinating and important sociological problem to identify what has given rise to it.

Whatever the reasons, idealism is currently extremely strong in sociology, and especially in the context of the sociology of knowledge. Much of the time, of course, it operates as no more than an epistemology of preference, providing a vocabularly for the formulation of issues that could equally adequately be formulated in materialist or even realist terms. But it does symbolize the decision of many sociologists in this area to narrow and differentiate their activities in a way which weakens their connections with a number of other fields, and it is perhaps the virtue of the (contrasting) approach taken here that it can offer an introduction to the subject which retains awareness of these connections.

It is perhaps also worth mentioning, before leaving readers to their own devices in this richly interesting field, that the idealism prevalent within and around it has strongly conditioned the way our own naturalistic approach has been 'read'. It has itself been extensively criticized (and occasionally supported) as an idealist sociological account which denies the existence of an external world and gives no role to experience in the generation of knowledge and belief. On this minor matter, this book could perhaps help to set the record straight. Induction from experience, however, suggests that it will not.[1]

Notes

1 A fine historical study of classification full of sociological interest is Winsor (1976). On the whole, however, studies of the natural sciences have given remarkably little attention to the direct word–world relation. In the philosophy of science Ian Hacking has lamented the tendency to speak not of observations but of observation statements: 'don't talk about things, talk about the way we talk about things' has been the rule in the field (1983, p. 167). And in the social sciences much the same trend is evident, even in some of the recent detailed studies of laboratory science. As a consequence science may appear as a form of abstract discourse rather than a practical activity: everything beyond the edge of speech, as it were, is left out of account. There are many reasons for the prevalence of this tendency, but it can easily arise inadvertently wherever academic fields rely upon the written word as their sole means of communication, and this same reliance makes difficulties for those who attempt to cure the problem (Gooding, Pinch and Schaffer, 1989; Gooding, 1990; Pickering, 1991).

2 This makes the connection with an earlier, seminal discussion by T.S. Kuhn (1974), which is considered in more detail in Barnes (1982a, Ch. 2; and 1984).

3 Again, reference to birds allows connection with a seminal example. Bulmer (1967) has discussed a New Guinea system of bird classification strikingly different from our own. It is compared with our system in Barnes (1984), where the two systems are claimed to stand on a par with each other as descriptions of nature, with nothing 'in nature' to indicate which 'corresponds to it' better.

4 This means that the account of applying a classification presently unfolding will also serve as an account of following a rule, and this will occasionally be presumed later. The *locus classicus* for this kind of understanding of rule-following is Wittgenstein (1968).

5 See Note 3 above.

6 Another discussion of the points to follow will be found in Barnes (1984).

7 Needless to say, the problem of how to determine death is a serious and absorbing one for medical practitioners, even though they apparently do it as a matter of course all the time. An intriguing recent incident is worth mention here. Every year many 'fatalities' in far-east Asia have been caused by imbibing a poisonous substance from inadequately prepared puffer-fish.

Apparently, these fish are such a delicacy that gourmets willingly take the risk of eating them. There has been a small but persistent movement of bodies from the restaurant to the crematorium. Recently, however, it has transpired that the secretions of the puffer-fish are used by voodoo doctors in Haiti to induce a state of catalepsy. The active ingredient is evidently related to chemicals which cause animal hibernation, and is capable of lowering bodily metabolism to well-nigh undetectable levels whilst maintaining full consciousness.

8 An extensive discussion of this 'problem of transitivity' and its epistemological importance will be found in Hesse (1974).

9 The argument now being developed follows that to be found in Barnes (1982a, Ch. 2, part 4). For an alternative way of making this kind of point, using the term 'know' in a different but equally tenable way, see Coulter (1989).

10 There is an important sociological project waiting to be done which would study just how it is possible for mathematics to be taught as it actually is taught.

11 Putnam's views have since changed, but his thoughts in his (1975), and especially in his long essay therein on 'The Meaning of "Meaning"', continue to be of enormous sociological interest.

12 In his later work this claim is perhaps slightly qualified. But, as we understand it, Collins' current position would still insist that a key part of *method* for sociological studies is acceptance of the postulate that reality has nothing to do with what is believed about it.

13 Needless to say, not all philosophical studies of induction are ruthlessly individualistic (Kornblith, 1985). Moreover, it is interesting to attempt to create links between Collins' sociological account and earlier approaches. In the context of the Bayesian approach to inductive inference, for example, Collins could be characterized as studying groups with strongly variant distributions of initial probabilities across their beliefs: the clash between parapsychologists and physicists could perhaps be modelled in Bayesian terms.

14 A very useful and insightful account of the relationship between inductive propensities in individuals and sociological studies of scientific reasoning is given by Nicholas (1984). It represents an ideal starting point for anyone who would seek fruitful connections between sociological attempts to understand induction as social activity and the longer-established approaches in the literature of philosophy and decision theory.

CHAPTER 4: BEYOND EXPERIENCE

1 The psychological literature on optical illusions referred to in Chapter 1 is of great technical interest and value as a contribution to the understanding of perception. It is also worthy of study as an example of the skillful yet unselfconscious use of a realist strategy – in that case the strategy which creates our sense of the illusory character of the various visual

perceptions. It is also interesting to attempt to imagine what would follow from an acceptance that the various illusions were not illusions at all: on the face of it, life would become more complicated but still practically possible.

2 In stressing on the one hand the universality of the realist mode of speech, and on the other the conventional character of all particular manifestations of it, our aim is to identify science as a form of culture like any other, and its own favoured modes of use of the realist mode of speech as amenable to sociological study. But these same points imply a *relativist* view of knowledge and culture which extends to sociology itself, and whatever knowledge claims it produces. This is a topic we ignore here, although it engenders enormous interest amongst sociologists and philosophers of science, and has generated an extensive literature. A recent collection in which a number of very different ways of dealing with this 'problem of reflexivity' are apparent is Pickering (1991). An extended overview of the relevant issues by a philosopher, and one which conveys the flavour of the relevant philosophical literature admirably, is Margolis (1986, 1987, 1989).

3 That this modest form of realism should be accepted in the sociological study of knowledge is argued in Barnes (1991), and indeed several workers in the field have no inhibitions about making reference to the state of the 'physical environment' or 'real world' or 'material world' as they attempt to account for changing knowledge. Others prefer to speak of 'resistance': they talk of accounts meeting resistance and not of their being difficult to reconcile with reality (Latour, 1987; Pickering and Stephonides, 1991). There is something to be said for this terminology, particularly since 'reality' is so easily confused with 'account of reality', and since knowledge claims are often alleged to be 'inconsistent with reality', when all that is meant is that they contradict currently accepted scientific knowledge. On the other hand, we need to keep in mind how much people everywhere achieve by use of common-sense realism, and by directly addressing their physical environment as something unverbalized and 'out there': this is not a trick recently invented by social scientists.

4 Although ethnomethodology is presented here as a means of studying realisms, it often presents itself as concerned with positivism, often thought of as the natural opposite of realism. Moreover, whereas ethnomethodology represents itself as strongly critical of positivism, to some readers it will appear itself as a form of positivism; in particular, its phenomenalism and its aversion to modelling and theorizing are typical features of positivism. These oddities, which need not affect a reading of the main text with its limited concerns, probably arise because 'positivism' is a term understood in a number of different ways, and with different denotations in different contexts.

5 It needs to be strongly emphasized that the point of the experiment was not to show that the 'answers' were irrelevant to subjects' definitions of the situation, or that they were incapable of redefining what was going on, or

that there is never any pattern in external experience save that pattern which is imposed upon it.

6 Although the main text implies that there need be no exercise of caution and restraint in the extension of this analogy, it may perhaps be that it is limited by considerations which are passed over in this book. These concern the extent to which bodies of knowledge have self-referring characteristics, and the way that the common-sense accounting practices of, for example, Agnes, 'happen as events in the same ordinary affairs that in organizing they describe' (Garfinkel, 1967, p. 1).

7 There are a number of different senses in which an observation may be said to be theory-laden, and they are so various and different in their implications that mere references to 'theory-ladenness' without further elaboration are well nigh meaningless. It is particularly unfortunate that references of this kind are now routinely used to imply that all observation reports are biased, or partial or unreliable. Hesse's (1974) discussion of theory and observation still stands as one of the clearest and most careful accounts of the relationship, and hence as an important resource for any student of the sociology of knowledge.

8 The connections between finitism in the context of the sociology of knowledge and finitist themes in ethnomethodology are nicely treated in Heritage (1984).

9 This is the key to understanding Whiggish modes of thought among natural scientists and the uniform tendency of scientist-historians to write Whig history. The requisite understanding of the honorific reward system in natural science was provided long ago in the seminal work of Robert Merton (1973).

10 Needless to say, this is no more a criticism of Olby than it is a criticism of the last few pages of this book. Textual exegesis is not pointless; what is odd about it is that its point surely cannot be what is generally taken to be its point.

11 The statistician R.A. Fisher, an extremely perceptive commentator on Mendel, noted that Mendel's empirical findings, far from being routinely replicable, are actually highly implausible and suggestive of systematically biased observation: they are far too close to the ideal segregation ratios to be plausible consequences of the natural operation of Mendelian segregation. A good modern repetition of Mendel's work would have to take care to achieve slightly inferior results if it were to count as confirmation of Mendel's views (Fisher, 1966).

12 For an earlier account which considers the finitist perspective upon theories at greater length, see Barnes (1982a).

13 There are immensely important issues lurking here, which merit a more extended discussion. Sets of exemplars do indeed make use of the same physical constants and the same empirical estimates of key quantities like the mass of the earth. But how far does the need to keep these constants fixed restrict the discretion of theorists and force them to proceed in one particular

way, so that the Duhem–Quine thesis becomes in practice irrelevant? And what is the nature of the need to keep these constants fixed: what kind of a transgression is it to refuse to keep them fixed?

A strictly finitist answer to the first question is that even after specific constants and values are accepted as fixed and non-negotiable, so many degrees of freedom will remain elsewhere in an exemplar that its further development will remain a matter of scientists' discretion. The task of keeping given constants and values fixed will merely increase the technical complexity of the task of extending an exemplar in one way rather than another. No fundamental change to the way we understand modelling will be entailed, no elimination of Duhem–Quine type problems.

A strictly sociological answer to the second question is that to accept the need to keep certain constants and values fixed is to align one's practice with the practice of other scientists, and nothing more. To share physical constants is to share conventions. To propose alternative values for the constants is to challenge the practice of other scientists. To propose the variability of such constants is to invite the community of scientists to accept and recognize additional forms of practice.

Whether these answers will stand up to detailed examination remains to be seen. There is much work to be done here. The approach of sociological finitism needs to be compared with that of philosophers (Sneed, 1979; Stegmüller, 1976). Some of the essays in Pickering (1991) attempt to press on further in the exploration of these problems.

14 The advantage of introducing the finitist account in relation to classification is that this observation-based form of activity is the one where the role of the physical environment is most clearly evident, and the powers of intervention of the scientist least obvious. In the last analysis, no distinction of any sociological significance exists between scientific practice in the observatory and in the laboratory.

CHAPTER 5: SOCIOLOGICAL PROJECTS

1 Brief reviews of the development of the sociological study of scientific knowledge can be found in Collins (1982, 1983): he lays less stress on the role of Kuhn's contribution, and emphasizes the significance of Wittgenstein's later philosophy, mainly as mediated through philosophical, anthropological and sociological writings. Our own resumé of sociological themes in Kuhn represents a particular way of reading his work (Barnes, 1982a), which involves a systematically different emphasis from that found in the historical and philosophical literature (Lakatos and Musgrave, 1970; Stegmüller, 1976; Gutting, 1980; Cedarbaum, 1983).

2 Pickering (1991) is intensely critical of traditional discipline-based approaches to the study of science, and argues that they need to be discarded and transcended if the full complexity of scientific practice and the many contingencies which affect it are properly to be appreciated. The article merits criticism in its turn for its impoverished conception of tradition,

expressed in a rhetoric which incorrectly assumes that any description which continues the tradition of a discipline must *ipso facto* be narrow, reductive and constraining. What Pickering attacks as narrow and selective accounting is actually that which highlights connections with very distant phenomena, thereby enriching the imagination, facilitating creative thinking, and calling attention to further aspects of the complexity of experience.

3 Perhaps it is wrong to imply that these are alternatives: possibly they could be made intelligible as aspects of a single underlying individualism. Certainly it would be an interesting intellectual exercise to attempt to produce such an account. But between the individualistic strand in ethnomethodology and Wittgensteinian philosophy on the one hand, and that in economics, rational decision theory and game theory on the other, there is, for the present at least, a striking discontinuity.

4 Thus, inspired by Latour, Rouse (1987) takes every scientific action without exception to be a political action, and conceives of the philosophy of science accordingly as political philosophy. The result is an extremely valuable reconceptualization of a number of philosophical and sociological problems, even if in the last analysis this particular presentation of the argument fails to make it stick: Rouse fails to avoid the problem of circularity referred to above in the main text.

5 It is actually most unusual for scholars who stress the nature of science as a form of life to take an interest in the explanation of scientific action. Generally they identify agreement in practice in a form of life and then speak of it no further, out of respect as it were (Lynch, 1991). Continuing the allusions of Note 3 above, the notion of explanation is taken to belong with the scientistic individualism of economics, decision theory and so forth, i.e. in the armoury of the opposite camp to those who speak of forms of life. Collins himself raises problems of explanation only rarely and with some reserve: it is possible that he has moved in this direction only reluctantly, explaining the massive agreement amongst scientists in 'social' terms in order to avoid being forced to treat it as a consequence of 'nature itself'; certainly this opposition is an enduring feature of Collins' thinking.

6 All the various myths which exaggerate the reliability of scientific knowledge, and encourage an unthinking trust in conclusions allegedly implied by it, have emerged precisely because of the enormous impact of actions and decisions of this kind (Smith and Wynne, 1989).

7 The relationship between the institution of responsible action and that of causal explanation is a conventional/reliable one. The false notion that the discourse of the two institutions is incompatible is encouraged by reference to inadequate accounts of causation, and idealized philosophical theories of what it involves, instead of to the actual characteristics of the discourse of cause and effect in practical contexts of use.

8 The view that there are no systematic distinctions to be drawn between the procedures of the social and the natural sciences is generally known as 'the thesis of naturalism'. Opponents of the thesis have sought to restrict

the scope of a 'scientific method' which they conceptualized in terms of the traditional stereotypes of science, and which, conceptualized in that way, they quite rightly reacted against as a way of understanding human behaviour. However, whilst they devoted all their energies to presenting the social sciences being pulled into the ambit of this rationalist conception of science, the natural sciences were themselves being reconceptualized. The thesis of naturalism became established by the very opposite kind of process to that which its opponents had sought to present. Adjustment to this change has been slow. Dualism is still sustained by a deep fear that the human and the social will otherwise be rationalized: that the assimilation could go the other way is not considered, still less that it is going the other way (cf. Margolis, 1986, 1987, 1989).

9 There are many other themes of sociological interest in this fascinating case study. The following discussion follows one strand of van den Belt's discussion exclusively.

10 Agreed, the judges might have tossed a coin. Indeed perhaps they did – in a back room. They would then have had to put on an appearance of 'rationality' by offering reasons in open court. The 'reasons' would consist in assertions that the sameness relations asserted on one side were more relevant or significant than those on the other. Given the lack of a metric for sameness, such reasons must presumably be *ex post facto* rationalizations.

11 Rice-Davies' law was itself of course first formulated in a court, as insightful commentary upon a witness's denial of Ms Rice-Davies' testimony that he had shared her bed.

12 Van den Belt's account of the role of the chemist A.W. Hoffman is exemplary material for consideration here. But as Hoffman was apparently less independent than he might have been, it serves our immediate purposes a little better to consider an imaginary witness.

13 It is necessary to decide on a pragmatic basis in every case how far such change needs to be taken into account. In van den Belt's study even the most directly relevant chemical knowledge rapidly changed. Little more than a decade after the initial verdict, the identification of the aniline reds as various salts of rosaniline was discarded: they 'proved to be complex and varying mixtures of various homologous and isomerous substances' (van den Belt, 1989, p. 196). And where relevant knowledge changes in this way, goals and interests are *ipso facto* liable to change as well.

How goals and interests change over time, and how these changes relate to changes in knowledge, are important sociological questions which must be passed over in this book. What can be stated, however, is that goals and interests, being features of social contexts and social relationships, will change much more rapidly than the superficially analogous characteristics which psychologists and biologists impute to individuals. See also Note 20 below.

14 The argument about nature which follows is given in a slightly more detailed version in Barnes (1987, p. 35 ff).

15 Recent history of science contains some marvellous studies of long-term scientific change in which the concerns of what were formerly the separate fields of social history and intellectual history are run together. The problem with taking a study like Hull (1988) or Rudwick (1985), now generally reckoned a classic of the genre, is that they are simply too rich in content and too demanding of space if they are to receive anything like appropriate treatment. Hence our use of some early work of Robert Kohler to rehearse some sociologically interesting issues. What follows, it must be emphasized, is not a critical commentary on Kohler's own views, which have in any case evolved since his papers of the early 1970s were written. In a more recent book (1982), he radically downgrades his earlier estimate of the role of zymase.

16 In his book-length study of the disciplinary politics of biochemistry, Kohler moves away from his earlier view that this manifesto had a key role in discipline formation, but it is debatable how far this earlier view is central to the argument of his original papers (Kohler, 1982).

17 Any discussion of extended historical change takes for granted substantial elements of what happened in order to concentrate on selected issues and episodes. In recounting such taken-for-granted background, the standard language of narrative of history of science is convenient and appropriate, since no reader will be misled into imagining that the episodes thus described are not themselves problematic and in need of further examination.

18 It may be useful, in considering this point, to recall the fourth and fifth tenets of finitism, as set out in Chapter 3.

19 The kind of process being described is familiar to historians of technological change, for whom the concept of 'advantages of adoption' is well exemplified (Arthur, 1984). When video recorders were first introduced different technologies were in competition. Buyers choosing between alternate formats would have initially been influenced by technical claims for their efficacy, even though all produced moving pictures. The different machines could be reconciled with the specifications of video reproduction, just as the different versions of zymase could be reconciled with Buchner's results, and yet technical argument about the machines and their adequacy mattered just as did arguments about different biochemical theories and their adequacy. But as decisions to buy were made, the technical advantages of the different video systems themselves changed. This was because the extent of use of a given system was actually one of the things which constituted its technical adequacy. The more a given kind of machine was bought, the cheaper it became per unit, the more its development and refinement could be financed, the greater the range of materials which would be produced for use on it, and so on. There are similar advantages related to adoption with scientific exemplars and theoretical vocabularies.

20 Shifting social alignments and changing goals and interests raise difficult yet crucially important questions for sociologists, and our intention here has been to acknowledge their existence rather than to embark on a serious

analysis of them. In the context of science they receive the most detailed attention in the work of Bruno Latour (1987, 1988), who concentrates particularly on what he refers to as the translation of interests: interests may be changed or translated, according to Latour, in so far as one actor can convince another of the expediency of the shift, but what remains problematic is how an actor is so convinced. Latour's ideas can be usefully compared with those of Hull (1988): both authors seek to understand the roles of power and expedient interest, which both recognize as ubiquitous in science, but they adopt very different explanatory schemes.

CHAPTER 6: DRAWING BOUNDARIES

1 Some historians of science have been surprised that Francis Bacon, advocate of the experimental method, was opposed to the methods of William Gilbert and Galileo, both of whom have been regarded as pioneers of the experimental method. Both men used experiments to test their preconceived hypotheses, however, and Bacon deplored this. William Harvey, another reknowned experimenter – but one who also subscribed to the hypothetico-deductive method of experimentation – dismissed Bacon's approach to the generation of knowledge. Bacon, he said, 'does philosophy like a Lord Chancellor'. By which he seems to have meant that Bacon tried to gather facts and witnesses to testify to those facts in the manner of a judge in a court of law.

2 Attempts to justify the claim that the 'problem of demarcation' is a genuine problem for the philosophy of science, which must be answered by pointing to some unique and distinctive aspect of scientific work, have met with little success (Gieryn, 1983; Mauskopf, 1990). The most well-known attempt to solve the problem is that of Karl Popper. Rejecting the 'verifiability' criterion of the logical positivists, Popper suggested 'falsifiability' as a criterion capable of separating science from non-science (Popper, 1968, pp. 40–4). Larry Laudan, another philosopher of science, has suggested, however, that Popper's criterion was specifically designed, not so much to help in reliably defining science but rather to ensure that what Popper took to be the overweening scientistic claims of Marxism and Freudianism were scotched (Laudan, 1983). The resulting demarcation criterion, as Laudan shows, allows what are currently considered to be clear cases of non-science to pass as science. More significant, from our point of view, is Laudan's acknowledgement that the mutability of the concept of 'scientific knowledge', as revealed by historical studies, defies any philosophical attempt to demarcate science from non-science once-and-for-all. Laudan has expressed despair about the possibility of reaching a philosophical solution to the problem of demarcation. This seems to be a conclusion hard to resist. The demarcation is always made as a result of social and historical contingencies and for pragmatic purposes and can only be understood in terms of historical sociology. For another sceptical view from a philosopher of science see Feyerabend (1975).

3 It was not part of Ospovat's own agenda to talk specifically about the drawing of boundaries in Victorian science but it is very easy to use his material for our purposes. All but one of the teleologists that Ospovat mentions are geologists: William Whewell, Charles Lyell, Adam Sedgwick, Henry de la Beche, William Buckland. The exception is the anatomist, Charles Bell. For the most part when he talks of the teleologists, Ospovat is talking, as he says himself, of 'the majority of British geologists' (p. 16), or 'most British geologists' (p. 14). The developmentalists, however, are all working in fields which we would now unhesitatingly designate as part of the biological sciences: Richard Owen, Peter Mark Roget, Martin Barry, William Carpenter, Louis Agassiz, Theodore Schwann, Alphonse De Candolle, Henry Milne Edwards, Ernst von Baer, Charles Darwin and 'other British biologists' (p. 10), or 'most of the leading naturalists' (p. 22) and 'the best young biologists' (p. 29).

4 Some readers may find this a fine distinction. While Victorian scientists could not legitimately invoke the direct action of God to explain physical phenomena, without abandoning their role as scientists, they were perfectly willing to concede that the laws of nature which they studied were established, imposed and maintained by God. Provided phenomena always behaved as if in accordance with laws, the scientist had a legitimate role in studying those phenomena and the governing laws. It is true to say, therefore, that for most Victorian scientists the concept of God was required to underwrite the existence of laws of nature – without God why should things behave in a lawlike way at all? This was not the same as explaining all things simply in terms of God's will.

5 The natural theologians were alarmed by the atheistic and politically subversive connotations which all evolutionary theories seemed to carry at this time (see Desmond, 1987, 1989). As Desmond shows, in spite of this opposition there were biologists and others who were willing to support evolutionary ideas.

6 Part of that evidence was the fossil record. Like the geologists, the biologists interpreted fossil remains as evidence for the radical discontinuity between creatures in different geological periods. The implication of this was that species remained the same for vast periods of time and were then wiped out and replaced by other fixed species. There was no evidence for one species changing into another. After the acceptance of evolutionary theories, following on from the publication of Darwin's *Origin of Species* (1859), the fossil record was reinterpreted and was seen to provide evidence of one species changing into another.

7 Perhaps it is worth adding here that geologists and biologists also shared a fence with the long-established and highly respected science of cosmological physics. The physicists insisted that the world could only be of a certain limited age because it was continually cooling down and its present temperature determined its maximum age. Geological evidence could only give relative ages, and so the geologists willingly used the

physicists' theories to enable them to make more absolute estimates, and in so doing gave further confirmation to the physicists' theories – and, therefore, their own! The physicists' estimates of the age of the earth were far too small to encompass the processes of evolution, and stood as empirical evidence against its validity. Darwin's later refusal to accept the validity of this seemingly debilitating criticism has been presented in some modern textbooks as a further tribute to his genius. His suggestion that not enough was known about the nature of the earth to be confident about the calculations of its age has been presented as a kind of prescience of the discovery of radioactivity which was to permit a greatly expanded age for the earth. But, at the time, this came close to a denial of the scientific credibility of thermodynamics, advanced purely in order to defend an unproven hypothesis. For many, biologists as well as physicists, it was Darwin whose scientific credentials were questionable, not the leading thermodynamicist, Lord Kelvin (Burchfield, 1975; Smith and Wise, 1989).

8 The literature is vast but see, for example, Bowler (1988, 1990). Ironically, the major advocate on behalf of uniformitarian geology in Britain, Charles Lyell, had turned his back on catastrophism, from about 1830, because he saw its progressionism as playing into the hands of those who wished to promote evolutionary theories in biology. Lyell's uniformitarianism was intended to give support to 'steady state' theories in geology, botany and zoology, and so undermine progressionist interpretations which he believed would lead inevitably to evolutionist conclusions (Bartholomew, 1973).

9 Historians and sociologists are generally in agreement in seeing boundary demarcation as an aspect of the ideology of science (e.g. Mulkay, 1976; Mendelsohn, 1977; Cooter, 1984; Gieryn, 1983; Barnes, 1985). Scientists present themselves and their disciplines in terms of a set of characteristics, ranging from abstract notions of methodology and intellectual values, to more concrete attributes such as institutional affiliation or representation (where was the work produced? where does this researcher work? where was the work first published? where has this researcher published previously? what experimental resources were used? and similar institutional considerations). The presentation usually serves to defend or enhance the professional integrity of science and its cognitive authority, which may in turn lead to other advantages (for example, in the competition for government funding).

10 An increasing number of historical studies indicates that there is little hope of demarcating science from non-science by looking at the procedures and arguments used by scientists themselves to draw up and maintain cognitive boundaries. Their rhetoric varies from one context to another, emphasizing Baconian induction here, the hypothetico-deductive method there, pointing to significant statistical correlation here while insisting upon absolute correlation there. In some cases genuine scientific knowledge claims are said to be obvious and undeniable to all, in others they are said to be understandable and confirmable only by an élite few (Shapin, 1990).

CHAPTER 7: PROOF AND SELF-EVIDENCE

1 There is an extensive and diverse literature devoted to the study of mathematical and logical reasoning from a naturalistic perspective. Some of this deals with developmental processes and some with studies of a more anthropological and descriptive kind. Among the former, of course, are the classic works by Piaget on the child's conception of number and reasoning, e.g. Piaget (1952). Among the latter see, for example, Gay and Cole (1967) and Cole and Scribner (1974). More recent attempts to penetrate the significance of mathematical concepts by attention to the situated or contextual character of their use are to be found in Livingston (1986), Lave (1988) and Watson (1990). For a critical discussion of Livingston's ethnomethodological approach see Bloor (1987). The approach adopted in the present chapter is mainly derived from Wittgenstein's *Remarks on the Foundations of Mathematics* (1978). For a further exploration and illustration of Wittgenstein's approach see Bloor (1983).

2 For a useful discussion of the background of the Frege–Russell approach, see Körner (1960).

3 The logical definition of the numbers in terms of sets of sets has not gone unchallenged. The difficulty is that there is more than one way of developing the theory of sets, and these lead to different ways of characterizing the numbers. In other words, the question: 'exactly *what* set is the number 1?' does not receive an unambiguous answer (see Benacerraf, 1965).

4 As well as high-brow and low-brow proofs there are also middle-brow proofs, e.g. those used by Frege and Peano that define the natural numbers in terms of the successor relation: 1 is the successor of 0, 2 the successor of 1 (and hence the successor of the successor of 0), etc. Symbolically $1 = S(0)$, $2 = S(1)$ or $2 = S(S(0))$. So $2 + 2 = S(S(0)) + S(S(0))$, and $4 = S(S(S(S(0))))$. Clearly, proving $2 + 2 = 4$ by using this apparatus requires us to keep track of the number of times 'S' is used. In effect it involves counting, and its employment thus presupposes the result it is designed to prove.

5 Calling *modus ponens* a truth table tautology means that 'if p then q, and p, therefore q' is true regardless of whether p and q are true. In this respect it is like the sentence, 'Either it is raining, or it is not raining', which is true regardless of the weather.

CONCLUSION

1 For extended reviews of the critical commentary relevant here, noting how misunderstanding is the typical feature of the literature in question, see Manicas and Rosenberg (1985, 1988) and Bloor (1991), 'Afterword', pp. 163–85.

Bibliography

Arthur, W.B. (1984) 'Competing technologies and economic prediction', *Options*, 2, 10–13.

Barnes, B. (1974) *Scientific Knowledge and Sociological Theory*, London, Routledge & Kegan Paul.

Barnes, B. (1976) 'Natural rationality: a neglected concept in the social sciences', *Philosophy of the Social Sciences*, 6, 115–26.

Barnes, B. (1977) *Interests and the Growth of Knowledge*, London, Routledge & Kegan Paul.

Barnes, B. (1982a) *T.S. Kuhn and Social Science*, London, Macmillan.

Barnes, B. (1982b) 'On the extensions of concepts and the growth of knowledge', *Sociological Review*, 30, 23–44.

Barnes, B. (1983) 'Social life as bootstrapped induction', *Sociology*, 17, 524–45.

Barnes, B. (1984) 'The conventional component in knowledge and cognition', in N. Stehr and V. Meja (eds), *Society and Knowledge*, New Brunswick, Transaction Books.

Barnes, B. (1985) *About Science*, Oxford, Blackwell.

Barnes, B. (1987) 'Concept application as social activity', *Critica*, 19, 19–44.

Barnes, B. (1991) 'Realism, relativism and finitism', in Raven and Thyssen (eds), *Cognitive Relativism and Society*, New Jersey, Rutgers University Press.

Barnes, B. and Bloor, D. (1982) 'Relativism, rationalism and the sociology of knowledge', pp. 21–47 in M. Hollis and S. Lukes (eds), *Rationality and Relativism*, Oxford, Blackwell.

Barnes, B. and Shapin, S. (eds) (1979) *Natural Order: Historical Studies of Scientific Culture*, London, Sage.

Bartholomew, M. (1973) 'Lyell and evolution: an account of Lyell's response to the prospect of an evolutionary ancestry for man', *British Journal for the History of Science*, 6, 261–303.

Belt, H. van den (1989) 'Action at a distance', in R. Smith and B. Wynne (eds), *Expert Evidence*, London, Routledge.

Benacerraf, P. (1965) 'What numbers could not be', *Philosophical Review*, 74, 47–73.

Ben-David, J. (1984) *The Scientist's Role in Society: A Comparative Study*, Chicago, University of Chicago Press.

Benveniste, J. (1988) '"High dilution" experiments a delusion: reply', *Nature*, 334, 291.

Black, M. (1962) *Models and Metaphors*, Ithaca, Cornell University Press.

Blackmore, J.T. (1972) *Ernst Mach: His Life, Work and Influence*, Berkeley, University of California Press.

Bloor, D. (1981) 'Hamilton and Peacock on the essence of algebra', pp. 202–32 in H. Mehrtens, H. Bos, and I. Schneider (eds), *Social History of Nineteenth-Century Mathematics*, Boston, Basel, Stuttgart, Birkhäuser.

Bloor, D. (1983) *Wittgenstein: A Social Theory of Knowledge*, London, Macmillan.

Bloor, D. (1987) 'The living foundations of mathematics' (review of E. Livingston, 1986), *Social Studies of Science*, 17, 337–58.

Bloor, D. (1991) *Knowledge and Social Imagery*, 2nd edn, Chicago, University of Chicago Press; (1st edn 1976, London, Routledge & Kegan Paul).

Bohr, N. (1913) 'On the constitution of atoms and molecules, pt.II', *Philosophical Magazine*, 26, 476–502.

Bowler, P. (1988) *The Non-Darwinian Revolution: Reinterpreting a Historical Myth*, Baltimore, Johns Hopkins University Press.

Bowler, P. (1990) *Charles Darwin: The Man and His Influence*, Oxford, Blackwell.

Boyer, C.R. (1959) *The History of the Calculus and its Conceptual Development*, New York, Dover.

Boyle, R. (1772) *Works*, 6 vols, London.

Brannigan, A. (1981) *The Social Basis of Scientific Discoveries*, Cambridge, Cambridge University Press.

Brewster, Sir David (1846) 'Explanations of vestiges of the natural history of creation', *North British Review*, 4, 504.

Broadbent, D.E. (1973) *In Defence of Empirical Psychology*, London, Methuen.

Bruner, J. and Postman, L. (1949) 'On the perception of incongruity: a paradigm', *Journal of Personality*, 18, 206–23.

Bulmer, R. (1967) 'Why is the cassowary not a bird?', *Man*, 2, 5–25.

Burchfield, J. (1975) *Lord Kelvin and the Age of the Earth*, New York, Science History Publications.

Campbell, D. (1979) *Descriptive Epistemology*, Cambridge, Mass., Harvard University Press.

Cartwright, N. (1983) *How the Laws of Physics Lie*, Oxford, Clarendon Press.

Cedarbaum, D.G. (1983) 'Paradigms', *Studies in the History & Philosophy of Science*, 14, 173–213.

Chambers, R. (1994a [1844]) *Vestiges of the Natural History of Creation and Other Evolutionary Writings*, edited by James A. Secord, Chicago, University of Chicago Press (1st edn of *Vestiges*, 1844, London, John Churchill).

Chambers, R. (1994b [1845]) *Explanations: A Sequel to the Vestiges of the Natural History of Creation*, London, John Churchill (included in Chambers, 1994).

Chambers, R. (1853) *Vestiges of the Natural History of Creation*, 10th edn, London, John Churchill.

Churchland, P. (1988) 'Perceptual plasticity and theoretical neutrality: a reply to Jerry Fodor', *Philosophy of Science*, 55, 167–87.

Clerk-Maxwell, J. (1873) *Treatise on Electricity and Magnetism*, Oxford, Clarendon Press.

Cole, M. and Scribner, S. (1974) *Culture and Thought: A Psychological Introduction*, New York, John Wiley.

Collingridge, D. and Reeve, C. (1986) *Science Speaks to Power*, London, Pinter.

Collins, H.M. (1974) 'The TEA set: tacit knowledge and scientific networks', *Science Studies*, 4, 165–86.

Collins, H.M. (1975) 'The seven sexes: a study in the sociology of a phenomenon, or the replication of experiments in physics', *Sociology*, 9, 205–24.

Collins, H.M. (1982) *Sociology of Scientific Knowledge*, Bath, Bath University Press.

Collins, H.M. (1983) 'The sociology of scientific knowledge: studies of contemporary science', *Annual Review of Sociology*, 9, 265–85.

Collins, H.M. (1992) *Changing Order: Replication and Induction in Scientific Practice*, London, Sage Publications.

Collins, H.M. and Pinch, T.J. (1982) *Frames of Meaning: The Social Construction of Extraordinary Science*, London, Routledge & Kegan Paul.

Cooter, R. (1984) *The Cultural Meaning of Popular Science: Phrenology and the Organization of Consent in Nineteenth-Century Britain*, Cambridge, Cambridge University Press.

Coulter, J. (1989) *Mind in Action*, Oxford, Polity Press.

Darwin, C. (1859) *On the Origin of Species by Means of Natural Selection*, London, John Murray.

Davenas, E., Beauvais, F., Benveniste, J., *et al.* (1988) 'Human basophil degranulation triggered by very dilute antiserum against IgE', *Nature*, 333, 816–18.

Davie, G. (1961) *The Democratic Intellect: Scotland and her Universities in the Nineteenth Century*, Part 2, 'The crisis in science', Edinburgh, Edinburgh University Press.

Dear, P. (1985) '*Totius in verba*: rhetoric and authority in the early Royal Society', *Isis*, 76, 145–61.

Desmond, A. (1987) 'Artisan resistance and evolution in Britain, 1819–1848', *Osiris*, 3, 77–110.

Desmond, A. (1989) *The Politics of Evolution: Morphology, Medicine and Reform in Radical London*, Chicago, University of Chicago Press.

Dirac, P.A.M. (1977) 'Ehrenhaft, the subelectron and the quark', in C. Weiner (ed.), *History of Twentieth-Century Physics*, New York, Academic Press, 1977, 290–3.

Duhem, P. (1914) *The Aim and Structure of Physical Theory*, trans. by P.P. Wiener, Princeton, NJ, Princeton University Press, 1954.

Durkheim, E. (1898 and 1974) 'Individual and collective representations', in E. Durkheim, *Sociology and Philosophy*, New York, the Free Press, ch. 1.

Durkheim, E. (1912) *The Elementary Forms of the Religious Life*, trans. by J.W. Swain, 1961, London, George Allen & Unwin.

Durkheim, E. and Mauss, M. (1908) *Primitive Classification*, trans. by R. Needham, 1963, London, Cohen & West.

Ehrenhaft, F. (1941) 'The microcoulomb experiment: charges smaller than the electronic charge', *Philosophy of Science*, 8, 403–57.

Engelhardt, H.T., Jr. and Caplan, A.L. (1987) *Scientific Controversies: Case Studies in the Resolution and Closure of Disputes in Science and Technology*, Cambridge, Cambridge University Press.

Enros, P. (1981) 'Cambridge University and the adaptation of analytics in early nineteenth-century England', in Mehrtens, Bos & Schneider (eds), *Social History of Nineteenth-Century Mathematics*, Boston, Basel and Stuttgart, Birkhäuser, pp. 135–48.

Farley, J. and Geison, G.L. (1974) 'Science, politics and spontaneous generation in nineteenth-century France: the Pasteur-Pouchet debate', *Bulletin of the History of Medicine*, 48, 161–98.

Felix, L. (1960) *The Modern Aspect of Mathematics*, trans. by J.H. and F.H. Hlavaty, New York, Basic Books.

Feyerabend, P. (1975) *Against Method: Outline of an Anarchistic Theory of Knowledge*, London, NLB.

Feynman, R.P., Leighton, R.B. and Sands, M. (1963) *The Feynman Lectures on Physics*, vol. 1, Reading, Mass., Addison-Wesley.

Fisher, R.A. (1966) 'Has Mendel's work been rediscovered?', in C. Stern and E.R. Sherwood (eds), *The Origins of Genetics: A Mendel Sourcebook*, San Francisco, Freeman.

Fodor, J.A. (1983) *The Modularity of Mind*, Cambridge, Mass., MIT Press.

Fodor, J.A. (1984) 'Observation reconsidered', *Philosophy of Science*, 51, 23–43.

Fodor, J.A. (1988) 'A reply to Churchland's "Perceptual plasticity and theoretical neutrality"', *Philosophy of Science*, 55, 188–94.

Forman, P. (1971) 'Weimar culture, causality and quantum theory, 1918–1927: adaptation by German physicists and mathematicians to a hostile intellectual environment', *Historical Studies in the Physical Sciences*, 3, 1–115.

Franklin, A. (1986) *The Neglect of Experiment*, Cambridge, Cambridge University Press.

Fuller, S. (1988) *Social Epistemology*, Bloomington and Indianapolis, Indiana University Press.

Garfinkel, H. (1967) *Studies in Ethnomethodology*, Englewood Cliffs, NJ, Prentice-Hall.

Gauquelin, M. (1984) *The Truth about Astrology*, Oxford, Blackwell.

Gay, J. and Cole, M. (1967) *The New Mathematics and an Old Culture: A Study of Learning among the Kpelle of Liberia*, New York, Holt, Rinehart & Winston.

Gibson, J.J. (1979) *The Ecological Approach to Visual Perception*, Boston, Howard Miflin.

Giere, R.N. (1988) *Explaining Science*, Chicago, University of Chicago Press.

Gieryn, T.F. (1983) 'Boundary-work and the demarcation of science from non-science: strains and interests in professional ideologies of scientists', *American Sociological Review*, 48, 781–95.

Gooding, D. (1990) *Experiment and the Making of Meaning*, Dordrecht, Kluwer Academic.

Gooding, D., Pinch, T. and Schaffer, S. (eds), (1989) *The Uses of Experiment*, Cambridge, Cambridge University Press.

Gregory, R. (1966) *Eye and Brian: The Psychology of Seeing*, London, Weidenfeld & Nicolson.

Gutting, G. (ed.) (1980) *Paradigms and Revolutions*, Notre Dame, University Press.

Haack, S. (1976) 'The justification of deduction', *Mind*, 85, 112–19.

Hacking, I. (1983) *Representing and Intervening: Introductory Topics in the Philosophy of the Natural Sciences*, Cambridge, Cambridge University Press.

Hanson, N.R. (1965) *Patterns of Discovery: An Enquiry into the Conceptual Foundations of Science*, Cambridge, Cambridge University Press.

Harre, R. (1972) *Philosophies of Science*, Oxford, Oxford University Press.

Harre, R. (1986) *Some Varieties of Realism*, Oxford, Basil Blackwell.

Heilbron, J.L. and Kuhn, T.S. (1969) 'The genesis of the Bohr atom', *Historical Studies in the Physical Sciences*, 1, 211–90.

Henry, J. (1988) 'The origins of modern science: Henry Oldenburg's contribution', *British Journal for the History of Science*, 21, 103–10.

Henry, J. (1990) 'Henry More versus Robert Boyle: the spirit of nature and the nature of providence', pp. 55–75 in S. Hutton (ed.), *Henry More (1614–1687): Tercentenary Studies*, Dordrecht, Kluwer Academic.

Henry, J. (1992) 'The scientific revolution in England', pp. 178–210 in R. Porter and M. Teich (eds), *The Scientific Revolution in National Context*, Cambridge, Cambridge University Press.

Heritage, J. (1984) *Garfinkel and Ethnomethodology*, Cambridge, Polity Press.

Hesse, M.B. (1974) *The Structure of Scientific Inference*, London, Macmillan.

Hobbes, T. (1651) *Leviathan: Or, The Matter, Form, and Power of a Commonwealth Ecclesiasticall and Civill*, London, Andrew Crooke.

Holton, G. (1978) 'Subelectrons, presuppositions and the Millikan–Ehrenhaft dispute', in *The Scientific Imagination*, Cambridge, Cambridge University Press, ch. 2.

Hull, D.L. (1988) *Science as a Process: An Evolutionary Account of the*

Social and Conceptual Development of Science, Chicago, University of Chicago Press.

Humphrey, G. (1963) *Thinking: An Introduction to its Experimental Psychology*, New York, Wiley.

Huxley, T.H. (1854) 'Vestiges of the Natural History of Creation', *British and Foreign Medico-Chirurgical Review*, 13, 425–39.

Kelvin, (Lord) (1897) 'Contact electricity and electrolysis according to Father Boscovich', *Nature*, 56, 1439, 84–5.

Keynes, J.N. (1906) *Studies and Exercises in Formal Logic*, London, Macmillan, 4th edn (first published 1884).

Kohler, R.E. (1971) 'The background to Eduard Buchner's discovery of cell-free fermentation', *Journal of the History of Biology*, 4, 35–61.

Kohler, R.E. (1972) 'The reception of Eduard Buchner's discovery of cell-free fermentation', *Journal of the History of Biology*, 5, 327–53.

Kohler, R.E. (1973) 'The enzyme theory and the origins of biochemistry', *Isis*, 64, 181–96.

Kohler, R.E. (1982) *From Medical Chemistry to Biochemistry*, Cambridge, Cambridge University Press.

Kornblith, H. (ed.) (1985) *Naturalizing Epistemology*, Cambridge, Mass., MIT Press.

Körner, S. (1960) *The Philosophy of Mathematics*, London, Hutchinson.

Kripke, S.A. (1972) 'Naming and necessity', in D. Davidson and G. Harman (eds), *Semantics of Natural Language*, Dordrecht, Reidel.

Kuhn, T.S. (1959) *The Copernican Revolution: Planetary Astronomy in the Development of Western Thought*, New York, Vantage Books.

Kuhn, T.S. (1961) 'Sadi Carnot and the Cagnard engine', *Isis*, 52, 367–74.

Kuhn, T.S. (1964) 'A function for thought experiments', *Mélanges Alexandre Koyré*, Paris, Hermann; reprinted in Kuhn (1977).

Kuhn, T.S. (1970) *The Structure of Scientific Revolutions*, Chicago, University of Chicago Press.

Kuhn, T.S. (1974) 'Second thoughts on paradigms', in F. Suppe (ed.), *The Structure of Scientific Theories*, Urbana, Ill., University of Illinois Press, 459–82; reprinted in Kuhn (1977), 293–319.

Kuhn, T.S. (1977) *The Essential Tension: Studies in Scientific Tradition and Change*, Chicago and London, University of Chicago Press.

Lakatos, I. (1971) 'History of science and its rational reconstructions', in R. Buck and R. Cohen (eds), *Boston Studies in the Philosophy of Science*, VIII, Dordrecht, Reidel.

Lakatos, I. (1976) *Proofs and Refutations: The Logic of Mathematical Discovery*, (eds J. Worrall & E. Zahar), Cambridge, Cambridge University Press.

Lakatos, I. and Musgrave, A. (eds) (1970) *Criticism and the Growth of Knowledge*, Cambridge, Cambridge University Press.

Latour, B. (1987) *Science in Action*, Milton Keynes, Open University Press.

Latour, B. (1988) *The Pasteurisation of France*, Cambridge, Mass., Harvard University Press.

Latour, B. (1993) *We Have Never Been Modern*, Hemel Hempstead, Harvester Wheatsheaf.

Laudan, L. (1977) *Progress and its Problems*, Berkeley, University of California Press.

Laudan, L. (1983) 'The demise of the demarcation problem', in R. Laudan (ed.), *Working Papers on the Demarcation of Science and Pseudo-Science*, Blacksburg, Virginia Tech. Center for the Study of Science in Society, 7–36.

Lave, J. (1988) *Cognition in Practice: Mind, Mathematics and Culture in Everyday Life*, Cambridge, Cambridge University Press.

Lewis, D.K. (1969) *Convention*, Cambridge, MA, Harvard University Press.

Livingston, E. (1986) *The Ethnomethodological Foundations of Mathematics*, London, Routledge & Kegan Paul.

Lynch, M. (1991) 'Extending Wittgenstein: the pivotal move from epistemology to sociology of science', in A. Pickering, *Science as Practice and Culture*, Chicago, University of Chicago Press.

MacKenzie, D.A. (1981) *Statistics in Britain 1865–1930. The Social Construction of Scientific Knowledge*, Edinburgh, Edinburgh University Press.

Mackie, J.L. (1966) 'Proof', *Proc. of the Aristotelean Society*, supp. vol. XL, 23–38.

Manicas, P.T. (1988) 'The sociology of scientific knowledge: can we ever get it straight?', *Journal for the Theory of Social Behaviour*, 18, 51–76.

Manicas, P.T. and Rosenberg, A. (1985) 'Naturalism, epistemological individualism and "The Strong Programme" in the sociology of knowledge', *Journal for the Theory of Social Behaviour*, 15, 76–101.

Mannheim, K. (1952) 'On the interpretation of Weltanschauung', in K. Mannheim, *Essays on the Sociology of Knowledge*, London, Routledge & Kegan Paul, 53–63.

Mannheim, K. (1952) *Essays on the Sociology of Knowledge*, London, Routledge & Kegan Paul.

Margolis, J. (1986) *Pragmatism without Foundations*, Oxford, Blackwell.

Margolis, J. (1987) *Science without Unity*, Oxford, Blackwell.

Margolis, J. (1989) *Texts without Reference*, Oxford, Blackwell.

Mauskopf, S.H. (1990) 'Marginal science', in R.C. Olby *et al.* (eds), *Companion to the History of Modern Science*, London, Routledge, 869–85.

Mendel, G. (1966) 'Experiments on plant hybrids', English translation of text of 1865 lecture, in C. Stern and E.R. Sherwood (eds), *The Origin of Genetics: A Mendel Sourcebook*, San Francisco, Freeman.

Mendelsohn, E. (1964) 'Explanations in nineteenth-century biology', in R. Cohen and M. Wartovsky (eds), *Boston Studies in the Philosophy of Science*, 2, Dordrecht, Reidel, 127–50.

Mendelsohn, E. (1977) 'The social construction of scientific knowledge', in E.

Mendelsohn, P. Weingart and R. Whitley (eds), *The Social Production of Scientific Knowledge*, Dordrecht, Reidel, 3–26.

Merton, R.K. (1973) *The Sociology of Science*, Chicago, University of Chicago Press.

Millar, R. (ed.) (1989) *Doing Science*, Lewes, Falmer Press.

Miller, H. (1870) *Footprints of the Creator: Or, the Asterolepis of Stromness*, 12th edn, Edinburgh, William P. Nimmo.

Millikan, R.A. (1913) 'On the elementary electrical charge and the avogadro constant', *Physical Review*, 2, 109–43.

Moore, G.N. (1982) *Zermelo's axiom of choice: its origins, development, and influence*, New York, Heidelberg and Berlin, Springer Verlag.

Moore, J. R. (ed.) (1989) *History, Humanity, and Evolution: Essays in Honour of John C. Greene*, Cambridge, Cambridge University Press.

Mulkay, M. (1976) 'Norms and ideology in science', *Social Science Information*, 15, 637–56.

Nelkin, D. (1975) 'The political impact of technical expertise', *Social Studies of Science*, 5, 35–54.

Nelkin, D. (ed.) (1979) *Controversy: Politics of Technical Decisions*, Beverly Hills, Sage.

Nicholas, J.M. (1984) 'Scientific and other Interests', in J.R. Brown (ed.), *Scientific Rationality: the Sociological Turn*, Dordrecht, Reidel, 265–94.

Nye, M.J. (1980) 'N–Rays. An episode in the history and psychology of science', *Historical Studies in the Physical Sciences*, 11, 125–56.

Olby, R. (1979) 'Mendel no Mendelian?', *History of Science*, 17, 53–72.

Olby, R.C., Cantor, G.N., Christie, J.R.R. and Hodge, M.J.S. (eds) (1990) *Companion to the History of Modern Science*, London, Routledge.

Oldenburg, H. (1996) *The Correspondence of Henry Oldenburg*, Volume III, edited by A.R. Hall and M.B. Hall, Madison, University of Wisconsin Press.

Orwell, G. (1949) *Nineteen Eighty-Four*, Harmondsworth, Penguin.

Ospovat, D. (1981) *The Development of Darwin's Theory: Natural History, Natural Theology, and Natural Selection, 1838–1859*, Cambridge, Cambridge University Press.

Paul, H.W. (1985) *From Knowledge to Power*, Cambridge, Cambridge University Press.

Piaget, J. (1952) *The Child's Conception of Number*, London, Routledge & Kegan Paul.

Pickering, A. (1980) 'The role of interests in high-energy physics: the choice between charm and colour', in K.D. Knorr, R. Krohn, and R. Whitley (eds), *The Social Process of Scientific Investigation*, Dordrecht, Reidel, 107–38.

Pickering, A. (1981) 'The Hunting of the Quark', *Isis*, 72, 216–36.

Pickering, A. (1984) *Constructing Quarks: A Sociological History of Particle Physics*, Edinburgh, Edinburgh University Press.

Pickering, A. (ed.) (1991) *Science as Practice and Culture*, Chicago, University of Chicago Press.

Pickering, A. and Stephanides, A. (1991) 'Constructing quaternions', in Pickering, A. (ed.), *Science as Practice and Culture*, Chicago, University of Chicago Press.

Pinch, T. (1986) *Confronting Nature: The Sociology of Solar-Neutrino Detection*, Reidel, Dordrecht.

Poincare, H. (1909) *Science and Method*, trans. F. Maitland, New York, Dover, n.d. (first published 1909: pagination from Dover edn).

Polya, G. (1954) *Mathematics and Plausible Reasoning*, 1: 'Induction and Analogy in Mathematics', Princeton, NJ, Princeton University Press.

Popper, K. (1968) *The Logic of Scientific Discovery*, 2nd edn, London, Hutchinson & Co.

Pumfrey, S. (1991) 'Ideas above his station: a social study of Hooke's curatorship of experiments', *History of Science*, 29, 1–44.

Purves, W.K. and Orians, G.H. (1987) *Life: the Science of Biology*, Sunderland, Mass., Sinawer.

Putnam, H. (1975) *Collected Papers*, 2, Cambridge, Cambridge University Press.

Reichenbach, H. (1947) *Elements of Symbolic Logic*, New York, Macmillan.

Richards, J. (1979) 'The reception of a mathematical theory: non-Euclidean geometry in England, 1868–1883', in B. Barnes and S. Shapin (eds), *Natural Order: Historical Studies of Scientific Culture*, London, Sage.

Richards, J. (1988) *Mathematical Visions: The Pursuit of Geometry in Victorian England*, London, Academic Press.

Rouse, J. (1987) *Knowledge and Power: Toward a Political Philosophy of Science*, Ithaca, Cornell University Press.

Rudwick, M.J.S. (1985) *The Great Devonian Controversy*, Chicago, University of Chicago Press.

Sainsbury, R. (1988) *Paradoxes*, Cambridge, Cambridge University Press, ch. 2.

Salmon, W.C. (1981) 'Rational prediction', *British Journal for the Philosophy of Science*, 32, 115–25.

Schaffer, S. (1989), 'Glass works: Newton's prisms and the uses of experiment', pp. 67–104 in D. Gooding, T. Pinch and S. Schaffer (eds), *The Uses of Experiment: Studies in the Natural Sciences*, Cambridge, Cambridge University Press.

Schelling, T.C. (1960) *The Strategy of Conflict*, Cambridge, MA, Harvard University Press.

Schutz, A. (1964) *Studies in Social Theory: Collected Papers*, 2, The Hague, Nijhoff.

Secord, J.A. (1989) 'Behind the veil: Robert Chambers and *Vestiges*', pp. 165–94 in J.R. Moore (ed.), *History, Humanity and Evolution: Essays for John C. Greene*, Cambridge, Cambridge University Press.

Sedgwick, A. (1850) *Discourse on the Studies of the University of Cambridge*, 5th edn, Cambridge, John Deighton.

Shapin, S. (1980) 'Social uses of science', in Rousseau and Porter (eds), *The Ferment of Knowledge: Studies in the Historiography of Eighteenth-Century Science*, Cambridge, Cambridge University Press, 93–139.

Shapin, S. (1982) 'History of science and its sociological reconstructions', *History of Science*, 20, 157–211.

Shapin, S. (1984) 'Pump and circumstance: Boyle's literary technology', *Social Studies of Science*, 14, 481–520.

Shapin, S. (1988) 'The house of experiment in seventeenth-century England', *Isis*, 79, 373–404.

Shapin, S. (1989) 'Who was Robert Hooke?', pp. 253–85 in M. Hunter and S. Schaffer (eds), *Robert Hooke: New Studies*, Woodbridge, Boydell Press.

Shapin, S. (1990) 'Science and the public', in R.C. Olby *et al.* (eds), *Companion to the History of Modern Science*, London, Routledge, 990–1007.

Shapin, S. and Barnes, B. (1979) 'Darwin and social Darwinism: purity and history', in Barnes and Shapin (eds), *Natural Order: Historical Studies of Scientific Culture*, London, Sage, 125–42.

Shapin, S. and Schaffer, S. (1985) *Leviathan and the Air-Pump: Hobbes, Boyle and the Experimental Life*, Princeton, Princeton University Press.

Shapin, S. and Thackray, A. (1974) 'Prosopography as a research tool in the history of science: the British scientific community, 1700–1900', *History of Science*, 12, 1–28.

Shapiro, B. (1983) *Probability and Certainty in Seventeenth-Century England: A Study of the Relationships between Natural Science, Religion, History, Law and Literature*, Princeton, Princeton University Press.

Smith, C. and Wise, N. (1989) *Energy and Empire: A Biographical Study of Lord Kelvin*, Cambridge, Cambridge University Press.

Smith, R. and Wynne, B. (eds) (1989) *Expert Evidence*, London, Routledge.

Sneed, J.D. (1979) *The Logical Structure of Mathematical Physics*, Dordrecht, Reidel.

Sprat, T. (1667) *The History of the Royal Society of London, for the Improving of Natural Knowledge*, London; reprinted in facsimile London, Routledge & Kegan Paul, 1959.

Stegmüller, W. (1976) *The Structure and Dynamics of Theories*, New York, Springer.

Stern, C. and Sherwood, E.R. (eds) (1966) *The Origin of Genetics: a Mendel Sourcebook*, San Francisco, Freeman.

Stouffer, S.A. (1949) *The American Soldier*, Princeton, Princeton University Press.

Thomason, B. (1982) *Making Sense of Reification*, London, Macmillan.

Tiles, M. (1989) *The Philosophy of Set Theory: An Introduction to Cantor's Paradise*, Oxford, Blackwell.

Tiles, M. (1991) *Mathematics and the Image of Reason*, London and New York, Routledge.

Turner, F. (1974) *Between Science and Religion: The Reaction to Scientific Naturalism in Late Victorian England*, New Haven, Yale University Press.

Turner, F. (1978) 'The Victorian conflict between science and religion: a professional dimension', *Isis*, 69, 356–76.

Waismann, F. (1982) *Lectures on the Philosophy of Mathematics*, Amsterdam, Rodopi.

Watson, H. (1990) 'Investigating the social foundations of mathematics: natural number in culturally diverse forms of life', *Social Studies of Science*, 20, 283–312.

Wertheimer, M. (1961) *Productive Thinking*, London, Tavistock.

Whewell, W. (1846) *Indications of the Creator*, 2nd edn, London, J. Parker.

Whitley, R. (1984) *The Intellectual and Social Organization of the Sciences*, Oxford, Clarendon Press.

Winsor, M.P. (1976) *Starfish, Jellyfish and the Order of Life*, New Haven, Yale University Press.

Wittgenstein, L. (1968) *Philosophical Investigations*, trans. by G.E.M. Anscombe, 2nd edn 1958, Oxford, Blackwell.

Wittgenstein, L. (1978) *Remarks on the Foundations of Mathematics*, trans. G.E.M. Anscombe, Oxford, Blackwell.

Wood, P.B. (1980) 'Methodology and apologetics: Thomas Sprat's *History of the Royal Society*', *The British Journal for the History of Science*, 13, 1–26.

Yeo, R. (1984) 'Science and intellectual authority in mid-nineteenth-century Britain: Robert Chambers and *Vestiges of the Natural History of Creation*', *Victorian Studies*, 28, 5–31.

Young, R.M. (1985) 'Malthus and the evolutionists: the common context of biological and social theory', in R.M. Young, *Darwin's Metaphor: Nature's Place in Victorian Culture*, Cambridge, Cambridge University Press, 23–25.

Index

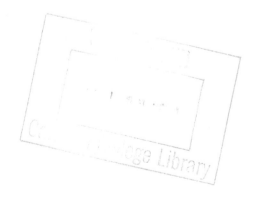